PUBLISHER COMMENTARY

We print NASA's handbooks and standards for the convenience of those that use them on a daily basis. We print all of these a full 8 ½ by 11 with large text so they are easy to read. Yes, color books are expensive to print so unless the information relies on the use of color for proper interpretation or understanding, we print most books in black and white to keep the cost down. All these documents are available for download for free from NASA, however printing them all over a network printer would take days.

Why buy a book you can download free? We print this so you don't have to.

All these books are available for free download from the government web site. Some are available only in electronic media. Some online docs are missing pages or barely legible.

We at 4th Watch Publishing are former government employees, so we know how government employees actually use the standards. When a new standard is released, an engineer prints it out, punches holes and puts it in a 3-ring binder. While this is not a big deal for a 5 or 10-page document, many NIST documents are over 100 pages and printing a large document is a time-consuming effort. So, an engineer that's paid $75 an hour is spending hours simply printing out the tools needed to do the job. That's time that could be better spent doing engineering. We publish these documents so engineers can focus on what they were hired to do – engineering. It's much more cost-effective to just order the latest version from Amazon.com

If there is a standard you would like published, let us know. Our web site is www.usgovpub.com

www.usgovpub.com

Copyright © 2019 4th Watch Publishing Co. All Rights Reserved

List of Other NASA Publications Available on Amazon.com:

NASA-STD-5001B	Structural Design and Test Factors of Safety for Spaceflight Hardware
NASA-STD-5006A	General Welding Requirements for Aerospace Materials
NASA-STD-5008B	Protective Coating of Carbon Steel, Stainless Steel, and Aluminum on Launch Structures, Facilities, and Ground Support Equipment
NASA-STD-5009A	Nondestructive Evaluation Requirements for Fracture-Critical Metallic Components
NASA-STD-5012B	Strength and Life Assessment Requirements for Liquid-Fueled Space Propulsion System Engines
NASA-STD-5019A	Fracture Control Requirements for Spaceflight Hardware
NASA-STD-5005D	Standard for The Design and Fabrication of Ground Support Equipment
NASA-HDBK-8739.21	Workmanship Manual for Electrostatic Discharge Control
NASA-HDBK 8739.23A	NASA Complex Electronics Handbook for Assurance Professionals (Color)
NASA-HDBK-8719.14	Handbook for Limiting Orbital Debris (Color)
NASA-HDBK-8709.22	Safety and Mission Assurance Acronyms, Abbreviations, and Definitions
NASA-HDBK-7009	NASA Handbook for Models and Simulations: An Implementation Guide For NASA-STD-7009 (Color)
NASA-HDBK-8739.19-2	Measuring and Test Equipment Specifications NASA Measurement Quality Assurance Handbook – Annex 2
NASA-HDBK-8739.19-3	Measurement Uncertainty Analysis Principles and Methods NASA Measurement Quality Assurance Handbook – Annex 3
NASA-HDBK-8739.19-4	Estimation and Evaluation of Measurement Decision Risk NASA Measurement Quality Assurance Handbook – Annex 4
NASA RCM	Reliability-Centered Maintenance Guide for Facilities and Collateral Equipment

www.usgovpub.com

Copyright © 2019 4th Watch Publishing Co. All Rights Reserved

NASA TECHNICAL HANDBOOK	NASA-HDBK-7004C
National Aeronautics and Space Administration Washington, DC 20546-0001	Approved: 11-30-2012 Superseding NASA-HDBK-7004B

FORCE LIMITED VIBRATION TESTING

MEASUREMENT SYSTEM IDENTIFICATION: METRIC (SI)/ENGLISH

APPROVED FOR PUBLIC RELEASE—DISTRIBUTION IS UNLIMITED

NASA-HDBK-7004C

DOCUMENT HISTORY LOG

Status	Document Revision	Approval Date	Description
Baseline		5-16-2000	Baseline Release
Revision	A	11-5-2002	Added New Section 6.0 "Comparison of Flight and Ground Vibration Test Data," which includes force data measured in two flight experiments. The flight data provide validation of the force limiting methodology. Made minor editorial changes.
Revision	B	1-31-2003	Page 8, Changed Equation 1(a) *from:* $f = f_o$ *to:* $f \geq f_o$ and Equation 1(b) *from:* $f = f_o$ *to:* $f \geq f_o$
Revision	C	11-30-2012	Reorganized document; added section 5.6 on impedance method; added section 6, Guidelines; added section 7.3 on GLAST spacecraft flight force data; moved material on Complex TDOF System and the spreadsheet example to Appendix C.

APPROVED FOR PUBLIC RELEASE—DISTRIBUTION IS UNLIMITED

NASA-HDBK-7004C

FOREWORD

This Handbook is published by the National Aeronautics and Space Administration (NASA) as a guidance document to provide engineering information; lessons learned; possible options to address technical issues; classification of similar items, materials, or processes; interpretative direction and techniques; and any other type of guidance information that may help the Government or its contractors in the design, construction, selection, management, support, or operation of systems, products, processes, or services.

This Handbook is approved for use by NASA Headquarters and NASA Centers, including Component Facilities and Technical and Service Support Centers. This Handbook establishes a common framework for consistent practices across NASA programs. This third revision of the Handbook includes several advances in the calculation and application of vibration force limits, guidelines for the application of force limiting, and data from a third flight experiment that involved measuring the forces and accelerations at the interface between a spacecraft and launch vehicle.

The primary goal of vibration tests of aerospace hardware is to identify problems that, if not remedied, would result in flight failures. This goal can best be met by implementing a realistic (flight-like) test with a specified positive margin. Usually, this goal is not well served by traditional acceleration-controlled vibration tests that historically do an adequate job of screening out flight failures but often result in failures that would not occur in flight. Vibration tests that are unrealistic and too severe are responsible not only for the cost and schedule overruns associated with hardware failures during unrealistic tests but also for the weight and performance penalties associated with designing for unrealistic tests.

It has been known for 40 years that the major cause of overtesting in aerospace vibration tests is associated with the large mechanical impedance of the shaker and the standard practice of controlling the input acceleration to the frequency envelope of the flight data (Blake, 1954 [1]; Salter, 1964 [2]; Murfin, 1968 [3]; Ratz, 1966 [4]; Heinricks, 1967 [5]; Painter, 1967 [6]). This approach results in unrealistic, large base reaction forces and other large responses at the fixed-base resonance frequencies of the test item. The conventional method of alleviating this problem is to measure and limit the acceleration responses in the test to those predicted for flight, but this approach is highly dependent on the analysis that the test is supposed to validate and usually requires limiting the acceleration responses at many locations on large test items.

This Handbook describes an improved vibration testing method that has been facilitated by two technological developments, circa 1980-1990: the advent of both three-axis piezoelectric force gages and shaker control systems with real-time response limiting. The subject method involves inserting the ring-shaped force gages under the test item to measure the reaction force between the test item and the shaker and limiting the measured force to a specification that is designed to replicate the flight environment. For most of the frequency range, the test is controlled, as in a conventional vibration test, by the input acceleration specification; but at the test item resonance frequencies, the force limit usually results in a reduction of the input acceleration (notching).

APPROVED FOR PUBLIC RELEASE—DISTRIBUTION IS UNLIMITED

NASA-HDBK-7004C

Since the mid-1990s, force limited vibration testing has been utilized on many projects, at the Jet Propulsion Laboratory (JPL), Goddard Space Flight Center (GSFC), other NASA Centers, and at other government laboratories and aerospace contractors. This Handbook describes the rationale and methodology of force limited vibration testing, so that the benefits of applying this relatively new technology can be maximized and the method can be applied throughout NASA in a consistent manner.

The Handbook is organized so that those already familiar with force limiting may use it as a reference and those who are new to this technology may use it as a textbook. In either case, it is not a Standard and therefore should be used only for guidance and not to justify a given approach or a specific numerical value. Section 6 of this Handbook gives three explicit guidelines based on years of experience conducting and overseeing force limited vibration tests.

Requests for information, corrections, or additions to this Handbook should be submitted via "Feedback" in the NASA Standards and Technical Assistance Resource Tool at http://standards.nasa.gov.

Original Signed By:
Michael G. Ryschkewitsch
NASA Chief Engineer

11-30-2012
Approval Date

APPROVED FOR PUBLIC RELEASE—DISTRIBUTION IS UNLIMITED

NASA-HDBK-7004C

TABLE OF CONTENTS

SECTION	**PAGE**
DOCUMENT HISTORY LOG	2
FOREWORD	3
TABLE OF CONTENTS	5
LIST OF FIGURES	6
LIST OF TABLES	7
1. SCOPE	8
1.1 Purpose	8
1.2 Applicability	8
1.3 Rationale	9
2. APPLICABLE DOCUMENTS	10
2.1 General	10
2.2 Government Documents	11
2.3 Non-Government Documents (References)	11
2.4 Order of Precedence	14
3. ACRONYMS AND DEFINITIONS	14
3.1 Acronyms and Abbreviations	14
3.2 Definitions	16
4. FORCE GAGES	17
4.1 Force Gage Characteristics	17
4.2 Force Gage Installation	18
4.3 Force Gage Preload	22
4.4 Force Gage Calibration	22
5. FORCE LIMITS	23
5.1 Response Limiting and Notching	24
5.2 Apparent, Effective, Residual, and Asymptotic Masses	25
5.3 Semi-Empirical Force Limit Equation	29
5.4 Simple TDOF System Method of Deriving Force Limits	31
5.5 Complex TDOF System Method of Deriving Force Limits	33
5.6 Equivalent Circuit Impedance Method of Deriving of Force Limits	33
5.7 Flight and Ground Test Scaling of C^2 in Semi-Empirical Method	37
5.8 FEM Analysis of Force Limits	38
5.9 Quasi-Static Load Limits	39
6. THREE GUIDELINES	40
6.1 Guideline 1: Use Force Limiting only for Highly Resonant Test Articles	40
6.2 Guideline 2: Use Appropriate Rationale for Deriving Force Limits	40
6.3 Guideline 3: Avoid Excessive Notching	41

APPROVED FOR PUBLIC RELEASE—DISTRIBUTION IS UNLIMITED

7.	**FLIGHT AND GROUND VIBRATION TEST DATA**	41
7.1	Shuttle Vibration Force (SVF-2) Experiment on Space Shuttle Space Transportation System-96 (STS-96)	41
7.2	Cosmic-Ray Isotope Spectrometer (CRIS) Instrument on Advanced Composition Explorer (ACE) Spacecraft	46
7.3	GLAST Spacecraft on Delta II Launch Vehicle	49

APPENDICES

A	Calculation of Effective Mass	55
B	Equations for Calculating the Simple TDOF System Force Limit	57
C	Tables for Complex TDOF System Method of Calculating Force Limits	58
D	Sweitzer's Notching Criterion	62

LIST OF FIGURES

FIGURE		PAGE
1	Deep Space One Spacecraft Mounted on 24 Small Force Transducers	19
2	Mars Exploration Rover Mounted on Eight Medium Force Transducers	20
3	CASSINI Spacecraft Mounted on Eight Large Force Transducers	21
4	Reduction of SDOF System RMS Response by Notching	25
5	Complex TDOF System with Residual and Modal Masses	27
6	Apparent Mass, Asymptotic Mass, Modal Mass, and Residual Mass of Longitudinally Vibrating Rod, Excited at One End and Free at Other End	28
7	Simple TDOF System of Coupled Oscillators	31
8	Normalized Force Specification from Simple TDOF System	32
9	Impedance Method Results for TDOF System with Two Identical Oscillators	35
10	SVF-2 Experiment on STS-96	42
11	Hitchhiker Canister for SVF-2	42
12	Force Limit in Vibration Test of SVF-2 Canister	43
13	Acceleration Input in Vibration Test of SVF-2 Canister	44
14	Y-Axis Acceleration at Top of SVF-2	45
15	Total Y-Axis Force in SVF-2 Flight	45
16	CRIS Instrument on ACE Spacecraft Bus	46
17	Total Vertical Force in CRIS Random Vibration Test	47
18	Notched Acceleration Input in CRIS Random Vibration Test	47
19	Spectral Density of Flight Normal Acceleration Measured near One Mounting Foot of CRIS Instrument	48
20	Spectral Density of Flight Normal Force Measured under CRIS Instrument	49
21	GLAST Spacecraft Sine Vibration Test	50
22	GLAST Spacecraft X-axis (Lateral) Interface Flight Accelerations	51
23	GLAST Spacecraft X-axis (Lateral) Base Flight Forces	51
24	GLAST Spacecraft Z-axis (Thrust) Interface Flight Accelerations	52
25	GLAST Spacecraft Z-axis (Thrust) Flight Base Forces	52

| 26 | GLAST Spacecraft Thrust Z-Axis Apparent Mass Measured in Ground Vibration Test and in Three Flight Regimes | 54 |
| 27 | GLAST Spacecraft Lateral X-Axis Apparent Mass Measured in Ground Vibration Test and in Two Flight Regimes | 54 |

LIST OF TABLES

TABLE		PAGE
1	Summary Table for C^2 Values in [29]	38
2	Force Limit Spectrum for Complex TDOF System with Q=50	59
3	Force Limit Spectrum for Complex TDOF System with Q=20	60
4	Force Limit Spectrum for Complex TDOF System with Q=5	61

NASA-HDBK-7004C

FORCE LIMITED VIBRATION TESTING

1. SCOPE

1.1 Purpose

The purpose of force limited vibration testing is to alleviate overtesting associated with the unrealistically high base reaction forces that occur at the test item resonances in conventional base-drive vibration tests (Blake, 1954 [1]; Salter, 1964 [2]; Murfin, 1968 [3]; Ratz, 1966 [4]; Heinricks, 1967 [5]; Painter, 1967 [6]); Smallwood, 1989 [7]; Scharton, et al., 1989 [8]; Smallwood, 1990 [9]; Scharton, 1995 [10]). The purpose of this Handbook is to provide an approach that may be consistently followed by those desiring to use force limiting, without having to conduct an extensive literature search or research and development effort before conducting the test. This Handbook describes the rationale behind force limiting, the hardware required for implementation, the various methods for deriving force-limit specifications, and three flight experiments in which the base reaction forces were measured.

The disparity in the mechanical impedances of the flight and the vibration test mounting configurations is usually the most important cause of overtesting in vibration tests. However, it may not be the only cause. A second contributor [2] to vibration overtesting is the uniformity of the motion at the various attachments of the test item to the vibration test fixture, which uniformity is often in contrast to the uncorrelated motion of the attachments in the flight-mounting configuration. The degree of overtesting associated with the uniformity of the attachment motion in the test is usually greater at the higher frequencies and when the flight-mounting configuration involves relatively wide-spread attachments to lightweight flight structure excited with an uncorrelated source, such as acoustic noise. Finally, overtesting can simply be the result of unrealistic test specifications, e.g., caused by the cascading of margins associated with the individual steps in the process used to develop the specifications. Neither of these latter two sources of overtesting is discussed further in this handbook.

1.2 Applicability

This Handbook is applicable to all force limited vibration tests of National Aeronautics and Space Administration (NASA) flight and non flight hardware, including but not limited to built-up spacecraft, spacecraft experiments and components, aircraft and launch vehicle equipment, launch vehicle pads, and Ground Support Equipment (GSE).

This Handbook is approved for use by NASA Headquarters and NASA Centers, including Component Facilities and Technical and Service Support Centers. This Handbook may also apply to the Jet Propulsion Laboratory (JPL) or to other contractors, grant recipients, or parties to agreements only to the extent specified or referenced in their contracts, grants, or agreements.

This Handbook, or portions thereof, may be referenced in contract, program, and other Agency documents for guidance. When this Handbook contains procedural or process requirements, they may be cited in contract, program, and other Agency documents for guidance.

APPROVED FOR PUBLIC RELEASE—DISTRIBUTION IS UNLIMITED

For the purpose of this Handbook, a force limited vibration test is any vibration test in which the reaction force at the base of the test item is measured and limited. In addition, a force limited vibration test almost always involves measuring and controlling the acceleration at the base of the test item and may also involve measuring and limiting the acceleration and/or force responses at other positions on the test item as well. For both analysis and test, it is usually convenient to think of the input acceleration as the control and of the base reaction force as a response, similar to the acceleration responses of the test item. The recommended means of measuring the base reaction force is with three-axis piezoelectric force gages, but other means, e.g., shaker armature current or strain gages, may be useful in special situations. Similarly, limiting the force is preferably implemented in real time by the shaker controller, but iterative, off-line control between runs may also be employed.

Since the purpose of force limiting is to reduce the base reaction force at test item resonances, the technique is most useful for highly resonant configurations, some examples of which are structure-like equipment such as telescopes, antennas, and reflectors; lightly damped hardware such as optics and cold stages; flight and development test model (DTM) spacecraft; and equipment with pronounced fundamental modes such as electronic boxes on flexible mounting. Therefore, force limiting is generally not needed in vibration tests of hard-mounted electronic boxes. Force limited vibration tests of various types of aerospace equipment are described in (Scharton, 1993 [11]; Chang and Scharton, 1996 [12]; Scharton and Chang, 1997 [13]; Scharton, 1998 [14]; Scharton, 2001 [15]; Scharton and Lee, 2003 [16]; and Soucy and Montminy, 2007 [17]). For example, force limited vibration tests of the Hubble Telescope 284-kg (625-lb) Wide Field Planetary Camera II (WFPCII), as well as of the 30-g (1-oz) Articulating Fold Mirror for the WFPCII, were conducted at NASA's JPL in the early 1990s [11]. The first of the three guidelines in section 6 of this Handbook is to use force limiting only for vibration tests of highly resonant systems and not for brick-like, non-resonant systems, such as hard-mounted electronic boxes.

1.3 Rationale

In base-drive vibration tests, where the shaker mechanical impedance is very high, the reaction force between the test item and the shaker can become very large at the fixed-base resonance frequencies of the test item. By contrast, in flight the subject equipment is typically mounted on relatively lightweight structure, which has a mechanical impedance comparable to that of the mounted equipment. In the flight mounting configuration, the support structure motion is reduced at the fixed-base resonance frequencies, which limits the interface forces, and as a result, extraordinarily large responses do not occur. This phenomenon is very similar to the behavior of a vibration absorber (Den Hartog, 1947 [18]). In the flight configuration, the resonances involve the combined motion of the equipment and its support, and these coupled-system resonances tend to have more damping and considerably less response than the fixed-base resonances of the equipment when it is mounted on a shaker. The purpose of force limiting is to reduce the response of the test item at its resonances on the shaker to more closely simulate the response of the combined system in the flight mounting configuration. The second guideline in section 6 of this Handbook is to take into consideration the flight mounting structural impedance in the determination of the amount of force limiting and notching to be employed in a force limited vibration test.

In most cases, only the total in-axis force, i.e., the sum of the forces at all mounting points in the shaker excitation axis, needs to be measured and limited in a force limited vibration test, although occasionally it is useful to measure the force at one or more individual mounting points, the cross-axis motion, and/or some moments. The amount of relief available from force limiting is greatest for highly resonant test items, which have large responses on the shaker and for test items that are mounted on relatively lightweight flight structures, which have mechanical impedances comparable to that of the test item. (The latter is typical.) Force limiting is most beneficial when the probability and/or penalties of a test failure are high. (Sometimes this is after an initial test failure.)

Force limiting should not be used to compensate for an acceleration specification that is thought to be too high, as the acceleration specification governs the response at both the resonance and the non-resonance frequencies. Also, most of the techniques discussed in section 5 of this Handbook for deriving force limits are based on the acceleration specification, so that if that is too high, the force limit will also be too high. Both the force and acceleration specifications should be appropriate envelopes, with the desired margin, of their respective flight environments at the interface between the test item and the mounting structure. Accordingly, the third and final guideline in section 6 of this Handbook counsels against misapplying force limiting as a means of reducing the broadband acceleration specification.

2. APPLICABLE DOCUMENTS

2.1 General

The documents listed in this section are applicable to the guidance in this Handbook.

2.1.1 The latest issuances of cited documents shall apply unless specific versions are designated.

2.1.2 Non-use of specific versions as designated shall be approved by the responsible Technical Authority.

The applicable documents are accessible via the NASA Standards and Technical Assistance Resource Tool at http://standards.nasa.gov or may be obtained directly from the Standards Developing Organizations or other document distributors.

2.2 Government Documents

NASA

NASA-STD-7001   Payload Vibroacoustic Test Criteria

2.3 Non-Government Documents (References)

1. Blake, R. E., "The Need to Control the Output Impedance of Vibration and Shock Machines," Shock and Vibration and Associated Environments, Bulletin No. 23, 1954.

2. Salter, J. P., "Taming the General-Purpose Vibration Test," Shock and Vibration and Associated Environments, Bulletin No. 33, Part III, 1964, pp. 211-217.

3. Murfin, W. B., "Dual Specifications in Vibration Testing," Shock and Vibration Bulletin, No. 38, Part 1, 1968, pp. 109-113.

4. Ratz, A. G., "An Impedance Compensated Random Equalizer," Proceedings of the Institute of Environmental Sciences 12th Annual Technical Meeting, San Diego, CA, April 1966, pp. 353-357.

5. Heinricks, J. A., "Feasibility of Force-Controlled Spacecraft Vibration Testing Using Notched Random Test Spectra," Shock and Vibration Bulletin, No. 36, Pt. 3, January 1967.

6. Painter, G. W., "Use of Force and Acceleration Measurements in Specifying and Monitoring Laboratory Vibration Tests," Shock and Vibration Bulletin, No. 36, Pt. 3, January 1967.

7. Smallwood, D. O., "An Analytical Study of a Vibration Test Method Using Extremal Control of Acceleration and Force," Proceedings of Institute of Environmental Sciences 35th Annual Technical Meeting, May 3, 1989, pp. 263-271.

8. Scharton, T. D., Boatman, D. J., and Kern, D. L., "Dual Control Vibration Testing," Proceedings of 60th Shock and Vibration Symposium, Vol. IV, 1989, pp. 199-217.

9. Smallwood, D. O., "Development of the Force Envelope for an Acceleration/Force Controlled Vibration Test," Proceedings of 61st Shock and Vibration Symposium, 1990, pp. 95-104.

10. Scharton, T. D., "Vibration-Test Force Limits Derived from Frequency-Shift Method," AIAA Journal of Spacecraft and Rockets, 32(2), 1995, pp. 312-316.

11. Scharton, T. D., "Force-Limited Vibration Tests at JPL—A Perfect Ten," ITEA Journal of Test and Evaluation, Vol. 14, No. 3, September 1993.

APPROVED FOR PUBLIC RELEASE—DISTRIBUTION IS UNLIMITED

12. Chang, K. Y. and Scharton, T. D., "Verification of Force and Acceleration Specifications for Random Vibration Tests of Cassini Spacecraft Equipment," Proceedings of ESA/CNES Conference on Spacecraft Structures, Materials & Mechanical Testing, Noordwijk, NL, March 1996.

13. Scharton, T. D. and Chang, K. Y., "Force Limited Vibration Testing of the CASSINI Spacecraft and Instruments," Proceedings of the Institute of Environmental Sciences, 43rd Annual Technical Meeting, 1997.

14. Scharton, T. D., "In-flight Measurements of Dynamic Force and Comparison with the Specifications Used for Limiting the Forces in Ground Vibration Tests," European Conference on Spacecraft, Structures, Materials, and Mechanical Testing, Braunschweig, GR, November 1998.

15. Scharton, T. D., "Force Limits Measured on a Space Shuttle Flight," Proceedings of the 47th Annual Technical Meeting of the Institute of Environmental Sciences and Technology, ESTECH 2001, April 22, 2001.

16. Scharton, T. and Lee, D., "Random Vibration Test of the Mars Rover Spacecraft," Proceedings of the 2003 Spacecraft and Launch Vehicle Dynamic Environments Workshop, Los Angeles, CA, June 2003.

17. Soucy, Y. and Montminy, S., "Investigation of Force Limited Vibration Based on Measurements During Spacecraft Vibration Testing," Proceedings of the 2007 Spacecraft and Launch Vehicle Dynamic Environments Workshop, Los Angeles, CA, June 2007.

18. Den Hartog, J. P., *Mechanical Vibrations*, 3rd Ed., McGraw-Hill, New York, 1947, p. 115.

19. Martini, K. H., "Multicomponent Dynamometers Using Quartz Crystals as Sensing Elements," ISA Transactions, Vol. 22, No. 1, 1983.

20. Kaufman, D., "Force Limiting for the Small Explorer Satellite Program at NASA Goddard Space Flight Center," 18th Aerospace Testing Seminar, March 1999.

21. O'Hara, G. J. and Remmers, G. M., "Measurement of a Structure's Modal Effective Mass," Shock and Vibration Bulletin, No. 39, Pt. 3, January 1969.

22. Wada, B. K., Bamford, R., and Garba, J. A., "Equivalent Spring-Mass System: A Physical Interpretation," Shock and Vibration Bulletin, No. 42, 1972, pp. 215-225.

23. Scharton, T. D., *Force Limited Vibration Testing Monograph,* NASA-RP-1403, May 1997, pp. 3-9.

24. Neubert, V. H., *Mechanical Impedance: Modeling/Analysis of Structures*, Jostens Printing and Publishing Co., State College, PA, 1987, p. 82.

25. Piersol, A. G., White, P. H., Wilby, E. G., and Wilby, J. F., "Vibration Test Procedures for Orbiter Sidewall Mounted Payloads -- Phase II Final Report," Astron Research and Engineering, Contract F04701-87-C-0010, USAF/AFSC, HQ Space Division, P.O. Box 92960, Los Angeles, CA 90009-2960, February 1, 1989, pp. 11-25.

26. Neubert, V. H., *Mechanical Impedance: Modeling/Analysis of Structures*, Jostens Printing and Publishing Co., State College, PA, 1987, Eq. 9.19.

27. Soucy, Y., Dharanipathi, V., and Sedaghati, R., "Comparison of Methods for Force Limited Vibration Testing," Proceedings of the IMAC XXIII Conference, Orlando, FL, January 31 – February 3, 2005.

28. Soucy, Y., Vijayamohan, V., and Sedaghati, R., "Investigation of Limit Criteria for Force Limited Vibration," Proceedings of the 2005 Spacecraft and Launch Vehicle Dynamic Environments Workshop, Los Angeles, CA, June 2005.

29. Soucy, Y., Dharanipathi, V., and Sedaghati, R., "Investigation of Force-Limited Vibration for Reduction of Overtesting," Journal of Spacecraft And Rockets, Vol. 43, No. 4, July–August 2006.

30. Staab, Lucas D., McNelis, Mark E., Akers, James, Suarez, Vicente J., and Jones, Trevor, M., *Application of the Semi-Empirical Force Limiting Approach for the CoNNeCT SCAN Testbed.* NASA-TM-2012-217627. (to be published)

31. Vujcich, M. and Scharton, T., "Combined Loads, Vibration, and Modal Testing of the QuikSCAT Spacecraft," Paper 1999-01-5551, AIAA and SAE 1999 World Aviation Conference, San Francisco, CA, October 1999.

32. Scharton, T., Pankow, D., and Sholl, M., "Extreme Peaks in Random Vibration Testing," Proceedings of the 2006 Spacecraft and Launch Vehicle Dynamics Environments Workshop, Los Angeles, CA, June 2006.

33. Kolaini, A. and Doty, B., "Statistical Analysis of Extreme Peaks in Random Vibration Response Data," Proceedings of the 2007 Spacecraft and Launch Vehicle Dynamics Environments Workshop, Los Angeles, CA, June 2007.

34. Miles, J. W., "On Structural Fatigue Under Random Loading," Journal of Aeronautical Sciences, November 1954, Eq. 3-5b.

35. Chang, K., "Structural Loads Prediction in Force-Limited Vibration Testing," Proceedings of the 2002 Spacecraft and Launch Vehicle Dynamics Environments Workshop, Los Angeles, CA, June 2002.

APPROVED FOR PUBLIC RELEASE—DISTRIBUTION IS UNLIMITED

36. Sweitzer, K. A., "A Mechanical Impedance Correction Technique for Vibration Tests," Proceedings of the Institute of Environmental Sciences 33rd Annual Technical Meeting, 1987, pp. 73–76.

37. Talapatra, D. C., McDonnell, R. H., and Hershfeld, D. J., "Analysis of STS-3 Get Away Special (GAS) Flight Data and Vibration Specifications for GAS Payloads," NASA Goddard Space Flight Center, Report 614-1, February 1983.

38. Gordon, S., and Kaufman, D., "Force Measurement on the GLAST Delta II Flight," Proceedings of the 2009 Spacecraft and Launch Vehicle Dynamics Environments Workshop, Los Angeles, CA, June 2009.

39. Gordon, S., Kaufman, D., Kern, D., and Scharton, T., "Prediction of Spacecraft Vibration Using Acceleration and Force Envelopes," Proceedings of the 2009 Spacecraft and Launch Vehicle Dynamics Environments Workshop, Los Angeles, CA, June 2009.

40. Delta II Payload Planners Guide, 06H0214, December 2006, p. 4-27.

41. NASA Engineering and Safety Center Technical Assessment Report: NESC-RP-06-071, September 2009 "Flight Force Measurements (FFMs) of the Gamma-Ray Large Area Space Telescope (GLAST) / Delta II Flight."

2.4 Order of Precedence

This Handbook provides guidance for the use of force limited vibration testing but does not supersede nor waive established Agency requirements/guidance found in other documentation.

3. ACRONYMS AND DEFINITIONS

3.1 Acronyms and Abbreviations

~	approximately
<	less than
%	percent
A	amplitude of the input acceleration specification
ACE	Advanced Composition Explorer
AFSC	Air Force Space Command
AIAA	American Institute of Aeronautics and Astronautics
CA	California
CG	center of gravity
CLA	coupled loads analysis
CNES	Centre Nation d'Etudes Spaciales
CoNNeCT	Communications, Navigation, and Networking reConfigurable Testbed
CRIS	Cosmic Ray Isotope Spectrometer
dB	decibel

DTM	development test model
ESA	European Space Agency
ESTEC	European Space Research and Technology Centre
F	force limit
FEA	finite element analysis
FEM	finite element model
FFM	Flight Force Measurement (project)
FFT	fast-Fourier transform
FL	Florida
FRF	frequency response function
G	gravity (text and figures)
g	gram(s)
g	gravity (figures)
GLAST	Gamma-Ray Large Area Space Telescope
GR	Germany
GRC	Glenn Research Center
GSE	ground support equipment
GSFC	Goddard Space Flight Center
HDBK	handbook
HH	Hitchhiker (canister)
HQ	headquarters
Hz	hertz
IMAC	International Modal Analysis Conference
in	inch(es)
ISA	International Society of Automation
ITEA	International Test and Evaluation Association
JPL	Jet Propulsion Laboratory
kg	kilogram(s)
lb	pound(s)
LVA	launch vehicle adapter
MATLAB	technical computing language (MATrix LABoratory)
MECO	main engine cut off
m	meter(s)
mm	millimeter(s)
NASA	National Aeronautics and Space Administration
NASTRAN	NASA Structural Analysis (system)
NESC	NASA Engineering and Safety Center
NL	Netherlands
oz	ounce(s)
PA	Pennsylvania
PAF	payload adapter fitting
PSD	power spectral density
Q	amplification; quality factor
QSLL	quasi-static load limit
QuikSCAT	Quick Scatterometer
RMS	root mean square

RP	reference publication
SAE	Society of Automotive Engineers
SCAN	Space Communications and Navigation
SDOF	single degree-of-freedom
SI	Système International
SRS	shock response spectra
STD	standard
STS	Space Transportation System
SVF-2	Shuttle Vibration Force (experiment) 2
TDOF	two degree-of-freedom
USAF	United States Air Force
WFPCII	Wide Field Planetary Camera II

3.2 Definitions

Acceleration of Center of Gravity (CG): Acceleration of instantaneous centroid of distributed masses, equal to external force divided by total mass.

Apparent Mass: Complex frequency response function (FRF) that is ratio of force to acceleration, a specific type of impedance.

C: Constant in semi-empirical method of deriving force limits.

Control System: The hardware and software that provide means for the test operator to translate vibration specifications into the drive signal for the shaker.

Dual Control: Control of both force and acceleration.

Dynamic Absorber: Single degree-of-freedom system tuned to excitation frequency to provide reaction force that reduces motion at attachment point.

Effective Modal Mass: The mass of a single degree-of-freedom system oscillator, which represents a mode of a vibratory system.

Extremal Control: Adjustment of vibration test input in each frequency band so that no limit channels exceed their specification.

Force Limiting: Reduction of the reaction forces in a vibration test to specified values, which are usually the interface forces predicted for flight, plus a desired margin.

Load: Test item.

Mechanical Impedance: Complex frequency response function that is the ratio of force to velocity, or the ratio of force to any motion quantity.

Notching: Reduction of acceleration input spectrum in narrow frequency bands, usually where test item has resonances.

Quality Factor (Q): The amplification (Q) of a single degree-of-freedom system at resonance.

Quasi-Static Acceleration: Combination of static and low-frequency loads into an equivalent load specified for design purposes as the CG acceleration.

Residual Mass: Sum of the effective masses of all modes with resonance frequencies greater than the subject frequency.

Response Limiting: Reduction of input acceleration to maintain measured response at or below specified value.

Shaker: The machine that provides vibratory motion to the test item, usually electro-dynamic, but can be hydraulic.

Single Degree-of-Freedom (SDOF) System: Vibration model with one mass attached to a base with a spring, sometimes called an oscillator.

Source: Test item support structure that provides flight excitation.

Tap Test: Measurement of force, acceleration, or both, while lightly striking a structure with a small instrumented hammer.

Test Fixture: Adapter that allows test item to be mounted to shaker.

Two Degree-of-Freedom (TDOF) System: Vibration model with two masses attached to a base with two or more springs

4. FORCE GAGES

Note: While the measurement system identification for this Handbook is, in general, Metric (SI)/English, certain units of measure included herein are traditionally and practically expressed in only one measurement system.

4.1 Force Gage Characteristics

The use of piezoelectric force gages for force limiting is highly recommended over other types of force measurement means, such as strain gages, armature current, etc. The advent of triaxial, piezoelectric, quartz force gages has made the measurement of force in vibration tests almost as convenient and accurate as the measurement of acceleration. Ring-shaped force gages, which are available in three or more sizes, are typically inserted around the bolts in the load path, and since these gages are very stiff, their presence has little effect on the dynamics of the test item. The high degree of linearity, dynamic range, rigidity, and stability of quartz make it an excellent transducer material for both accelerometers and force gages (Martini, 1983 [19]). (Single-axis force gages are also available, and some of these use other piezoelectric materials, which have higher sensitivities than quartz but typically are not as stiff nor as linear. Also, the use of triaxial gages provides the capability to measure and limit the cross-axis motion and often eliminates the need to reinstall the gages when the test axis is changed.)

Similar signal processing and charge and voltage amplifiers may be used for piezoelectric force gages and accelerometers. Force gages, like accelerometers, are available either with or without integral charge amplifiers. The advantage of using gages without integral charge amplifiers is that the sum of the forces measured by the gages at all the mounting points may be obtained by using a simple junction box to sum the charges from all the gages before they are converted to voltage. Alternately, if force gages with integral charge amplifiers are used, a special voltage-summing box available from the gage manufacturers has to be employed to provide the total force. Although more complex and expensive, the use of a voltage-summing box does provide the capability for measuring both moments and forces at the individual mounting points, together with the sum of the forces. Finally, piezoelectric force gages tend to put out more charge than piezoelectric accelerometers because the force gage crystals experience higher loading forces, so sometimes it is necessary to use a charge attenuator before the charge amplifier(s).

4.2   Force Gage Installation

The preferred method of installing the force gages is to sandwich one gage between the test item and conventional vibration test fixture at each test item attachment position and to use fasteners that are longer than the flight fasteners to accommodate the thickness of the gages. In this configuration, there is no fixture weight above the transducers, and the force acting on the gage is nearly equal to the force acting on the test item. (The preload bolt also carries some of the force; see section 4.3 of this Handbook.) Sometimes, however, this preferred approach is impractical, e.g., if the attachments involve shear pins or groupings of closely spaced bolts. In these cases, it may be necessary to use one or more adapter fixtures to interface the transducers to the test item. A good rule of thumb is that the total weight of the adapter fixtures above the force gages should not exceed 10 percent of the weight of the test item. This limitation is necessary because the force gages read the sum of the force required to accelerate the adapter fixture(s) and that delivered to the test item. Therefore, if a heavy adapter is utilized, the force necessary to move the adapter, i.e., the mass of the adapter times the input acceleration specification, should be added to the force limit derived for the test item itself. If the adapter fixture weight exceeds the 10 percent criterion, force limiting will usually be useful only for the first test item vibration mode in each axis. At higher frequencies, the force consumed by the adapter will result in a noise floor, which masks measurement of the smaller forces associated with the higher modes of the test item. Use of a circuit to subtract the adapter fixture force in real time has been tried, but it is not recommended because of the phase errors that result when the adapter fixture installation is not perfectly rigid. The use of armature current to measure shaker force is also not recommended, because the weight of the shaker armature and fixtures typically are much greater than 10 percent of the test item weight and also because of phase errors associated with electromechanical resonances.

It is also recommended that the test fixture be stiff in the frequency range of force limiting application to avoid fixture resonance reactions, and the entire load bearing areas on both sides of the force transducers have to be engaged to prevent force transducer damage and to obtain an accurate force reading. The addition of precision washers may be necessary in some installations to provide this interface at the force transducer.

Figure 1, Deep Space One Spacecraft Mounted on 24 Small Force Transducers; figure 2, Mars Exploration Rover Mounted on Eight Medium Force Transducers; and figure 3, Cassini Spacecraft Mounted on Eight Large Force Transducers, show three examples of force gage and adapter fixture installations used for force limited vibration tests of large test items. Figure 1 shows the Deep Space One spacecraft tested with 24 small size, 8.1-mm (0.319-in) inside diameter, 225-kg (550-lb) axial force capability, triaxial force transducers, which are positioned around each of the 24 bolts with which the spacecraft attaches to the launch vehicle adapter. Figure 2 shows the Mars Exploration Rover and base petal mounted on eight medium size, 26.5-mm (1.043-in) inside diameter, 2,022-kg (4,400-lb) axial force capability, triaxial force transducers, which are used in pairs with special adapter fixtures that preload the gages and mate to the Rover base petal. Figure 3 shows the Cassini spacecraft tested on eight large, 40.5-mm (1.594-in) inside diameter, 8,136-kg (17,900-lb) axial force capability, triaxial force gages, which are mounted under a 91-kg (200-lb) adapter ring to which the spacecraft launch vehicle adapter is bolted.

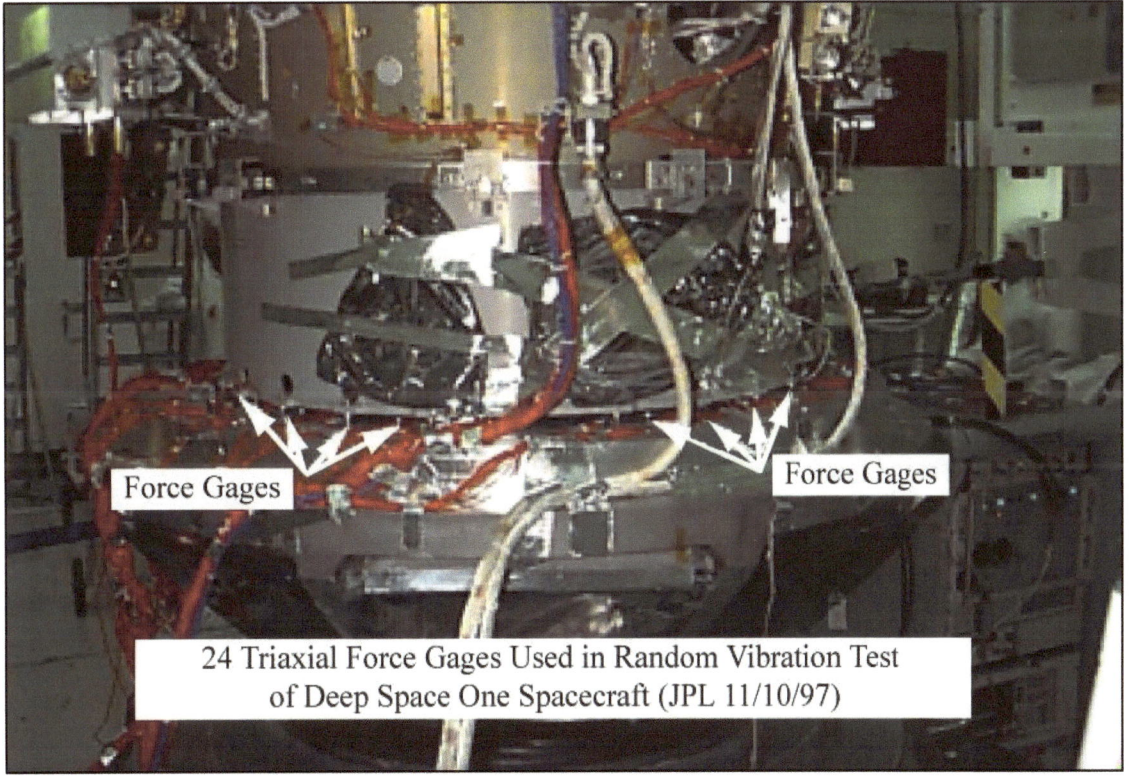

Figure 1—Deep Space One Spacecraft Mounted on 24 Small Force Transducers

Figure 2—Mars Exploration Rover Mounted on Eight Medium Force Transducers

Figure 3—Cassini Spacecraft Mounted on Eight Large Force Transducers

4.3 Force Gage Preload

Force gages have to be inserted between (in series with) the test item and shaker, and the gages have to be preloaded so that the transducer always operates in compression. The preload is achieved using a threaded bolt or stud, which passes through the inside diameter of the transducer ring. The static compression force in the transducer is balanced by the static tension in the bolt. As described in section 4.2 in this Handbook, using flight hardware with longer fasteners to accommodate the gage is the simplest and preferable mounting configuration. However, with this configuration, it is usually impossible to achieve the force gage manufacturer's recommended preload values, which are typically very high. (Having a high preload, as well as smooth transducer and mating surfaces, is advantageous because it minimizes several common types of measurement errors, e.g., bending moments being falsely sensed as tension/compression.) Usually, however, the recommended preload values are greater than the flight preloads; in addition, sometimes it is necessary to trade off transducer preload and dynamic load carrying capability, particularly for moments. The three requirements for selecting the preload are (1) that the maximum stress on the transducers does not exceed that associated with the maximum load set specified by the manufacturer; (2) that the preload is sufficient to carry the shear loads via friction, without slip; and (3) that the preload is sufficient to prevent unloading related to the dynamic forces and moments, e.g., tensile forces and heel-to-toe moments.

When the bolt preload is critical, which may be the case for large test items like spacecraft, the force transducers may be used to measure the bolt preload while the bolts are being torqued. (The actual preload resulting from a specified value of torque may vary by a factor of two or more, depending on friction and lubrication.) Piezoelectric force transducers are dynamic, not static sensors; however, when charge amplifiers with high input resistance and a long time constant setting are utilized, the preload force readings for each gage will hold steady for hours, which is ample time to complete the bolt torquing sequence.

4.4 Force Gage Calibration

The force gage manufacturer typically provides a nominal calibration for each transducer, but the sensitivity of installed units depends on the preloading configuration, because the preload bolt carries a portion of the dynamic load. The sensitivity of the force gage itself does not depend significantly on the amount of preload but only on the preloading configuration. The transducer and the bolt act like two springs in parallel, and the fluctuating load carried by each is proportional to its stiffness. Therefore, the sensitivity (charge or voltage divided by force) of the installed configuration is equal to the sensitivity of the transducer itself multiplied by the ratio of the transducer stiffness to the sum of the bolt plus the transducer stiffnesses. (The flexibility of any structural elements, mounting feet, etc., in these two load paths also have to be included in this calculation.) The transducer stiffness is available from the manufacturer, and the stiffness of the bolt and structural elements can be estimated from strength-of-materials or finite-element-model (FEM) calculations. Sometimes the transducer manufacturers provide two sets of calibration values — one for the transducer itself and one for the transducer with their standard preloading hardware. (If they only provide calibration values with their standard preload configuration, it may be necessary to back out the effect of the preload bolt using the above considerations.) (Although some transducer manufacturers offer standard preload hardware, for aerospace applications it is usually preferable to utilize a preloaded bolt configuration that is tailored to the test item, as described in section 4.2 of this Handbook.)

In situ calibration of force gages may be accomplished either statically or dynamically. The easiest method of calibrating the transducers for a force limited vibration test is to conduct a preliminary low-level sine sweep or random run. The low-frequency asymptote of the apparent mass is compared with the known total mass of the test item. (The relevant apparent mass is the ratio of total force to the input acceleration in the shake direction.) If it is not possible to start the sweep at a frequency sufficiently below the first resonance of the test item to avoid significant amplification, it may be necessary to use the single degree-of-freedom (SDOF) system transmissibility curve to determine a correction factor to the low-frequency asymptote. Typically, the measured force (before correcting for the preload bolt), using the manufacturer's calibration number for the transducer itself, will be approximately 80 to 90 percent of the test item and above-gage fixturing total weight in the axial direction and 90 to 95 percent of the total weight in the lateral directions, where the preloading bolt is usually less stiff in shear or bending.

If the low-frequency asymptote of apparent mass is not above the aforementioned values and the discrepancy cannot be explained by the stiffness of the preloading bolt or another shunt path, then the test should be stopped until the problem is resolved. Usually, the discrepancy will be related to a problem with the force gage cabling or connections or to an error in the charge or voltage amplifier settings. Less often, it may be related to cross-axis coupling or an error in the assumed physical mass of the test item and/or fixture above the gages.

The transducer installation may also be calibrated statically, using weights or a load cell with the charge amplifier configuration discussed in section 4.3 of this Handbook for measuring preload. If weights are used, it is recommended that the calibration be performed by first loading the transducers, zeroing out the charge, and then removing the load, to minimize any transient overshoot associated with the load application. If the transducers are not preloaded during the static calibration, it is important to apply sufficient weight or force to overcome any gaps between the individual quartz elements.

Additional information on the physics, limitations, installation, preload, and calibration of piezo-electric force gages is contained in [19] and in the gage manufacturer's data sheets.

5. FORCE LIMITS

Force limits are analogous and complementary to the acceleration specifications used in conventional vibration testing. Just as the acceleration specification is the frequency envelope of the flight acceleration at the interface between the test item and flight mounting structure, the force limit is the envelope of the flight total force at that interface. In force limited vibration tests, both the acceleration and force specifications are needed, and the force specification is generally based on and proportional to the acceleration specification. Therefore, force limiting does not compensate for errors in the development of the acceleration specification, e.g., too much conservatism or the lack thereof. These errors will carry over into the force specification. Force limits are usually derived from coupled-system analyses and impedance (apparent mass) information obtained from tests or finite element analysis (FEA). Also, interface forces have been measured in several flight experiments, three of which are described in section 7 of this Handbook, and a semi-empirical method of predicting force limits is described in section 5.3 of this Handbook.

Force limits are typically developed in one-third octave bands, and it is recommended to specify the force limit as a smooth, slowly varying curve, much as one specifies acceleration inputs. Force spectra typically roll off with frequency much faster than do acceleration spectra. Therefore, it is usually adequate to specify the force limits only in the frequency regime encompassing the first few modes in each axis, which might be out to approximately 100 Hz for a large spacecraft, 500 Hz for an instrument, or 2,000 Hz for a small component. However, it is important to recognize that the test item resonances on the shaker occur at considerably higher frequencies than in flight. Therefore, care should be taken not to roll off the force specification at a frequency lower than the fundamental resonances on the shaker. In general, the rolloff of the force specification should follow the rolloff of the asymptotic mass. (See section 5.3 of this Handbook.)

This section discusses response limiting in general and the resulting notching of the input, the concept of apparent mass, a simple two degree-of-freedom (TDOF) model that provides a basic understanding of the physics of force limiting, and several other methods of deriving force limits. The relation between the interface force and acceleration and a survey of historical methods of deriving force limits are discussed in (Kaufman, 1999 [20]), where the methods are categorized according to whether they depend on the source, the load, or both. Herein, the emphasis is on methods that depend on both the source and load and specifically on the ratio of the load to source apparent mass at the fixed-base resonance frequencies of the test item, because this ratio is believed to be the key parameter needed to evaluate the force limit and resultant notching.

5.1　Response Limiting and Notching

In a vibration test, it is usually convenient to think of the acceleration specification as the input, or control, and the base reaction force as a response, similar to the acceleration responses measured on the test item. In addition to controlling the test to the input acceleration specification, most modern vibration test controllers have the capability to limit a number of response channels to separate specifications. Thus, at frequencies where one of these response limits would be exceeded if the input acceleration specification were followed, control shifts to that response channel, and the input acceleration is reduced, i.e., notched. This vibration test controller capability is commonly referred to as extremal control. It is in this manner that force limiting is usually implemented.

The depth of a notch in a force limited vibration test and in any other response-limited test depends on two things: the force (or response) limit and the damping of the resonance being limited in the test. When the flight mounting structure and test item impedances are comparable, the force or response limit is relatively insensitive to the test item damping, but the damping determines how much the force or response in the test would exceed the limit and thus the depth of the notch that results from limiting. Therefore, lightly damped resonances are notched much deeper than heavily damped ones. (See section 5.4 of this Handbook.)

Notching the acceleration input using force or acceleration response limiting is generally not as effective in reducing the root mean square (RMS) response as is reducing the acceleration input specification at all frequencies. Figure 4, Reduction of SDOF System Mean-Square Response by Notching, shows the reduction of the RMS response of an SDOF system resulting from notching. For example, a 14-dB notch reduces the mean-square response by a factor of four and thus the RMS response by only a factor of two, or 6 dB; and even a notch of 20 dB or more will not reduce the RMS response by more than a factor of three, or 10 dB.

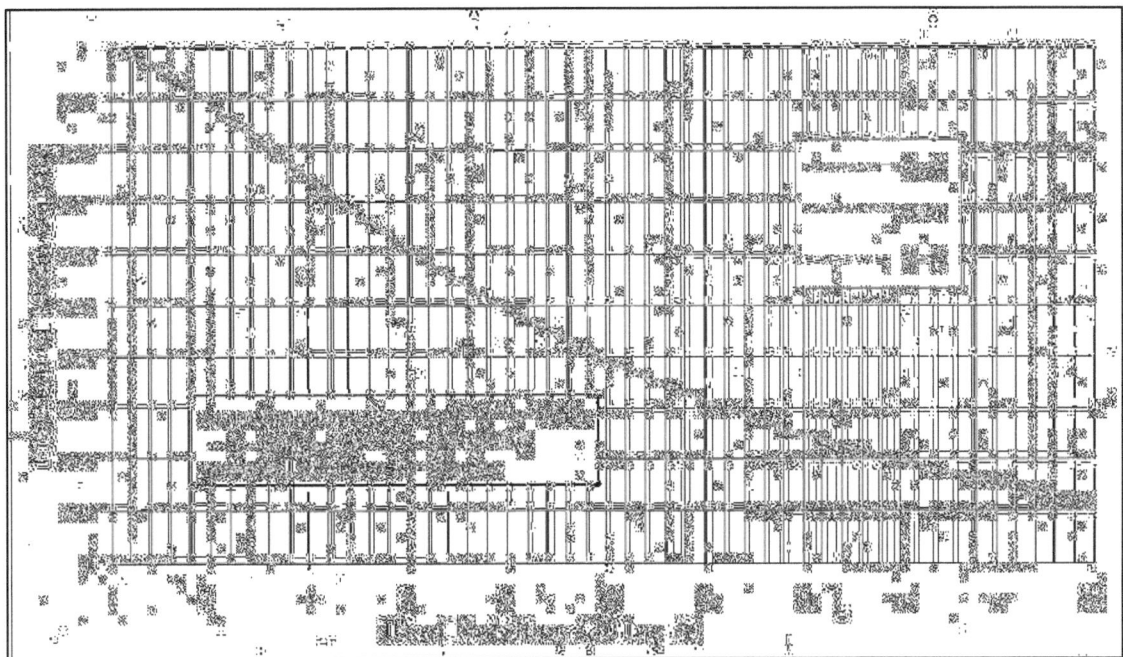

Figure 4—Reduction of SDOF System Mean-Square Response by Notching

5.2 Apparent, Effective, Residual, and Asymptotic Masses

The rationale for force limiting discussed in section 1.3 of this Handbook is based on the disparity between the impedances of typical aerospace mounting structures and the large impedances of vibration test shakers. Therefore, the derivation of force limits is based primarily on a consideration of the mechanical impedance of the flight mounting structure and specifically on the ratio of the test item mechanical impedance to that of the flight mounting structure. Mechanical impedance is usually defined as the ratio of force to velocity, but herein it is preferred to use the frequency response function (FRF) (the ratio of force to acceleration), which is usually called apparent mass.

The concept of effective mass (O'Hara and Remmers, 1969 [21] and Wada, et al., 1972 [22]) decomposes the physical mass of an object into its modal components. Consider the drive point apparent mass of the model consisting of the set of SDOF systems connected in parallel to a rigid, massless base as shown in figure 5a, Complex TDOF System with Residual and Modal Masses (Asparagus Patch Model of Source or Load) [10]. The masses of the SDOF systems connected in parallel to the drive base in figure 5a are the effective masses of the modes represented by the SDOF systems. The sum of the modal effective masses for each excitation axis is equal to the total mass of the distributed system. The sum of the effective masses of the modes with resonance frequencies above the excitation frequency is called the residual mass. Figure 5b, Residual and Modal Mass Model of Source or Load, illustrates a model of a single mode with both modal and residual masses. Figure 5c, Coupled TDOF System Residual and Modal Mass Model, shows a model of two such modes coupled together, which, ignoring the zero frequency mode, is a rather complex TDOF system that is analyzed in [10]. Effective modal and residual masses are typically calculated when FEM codes, such as NASTRAN, are used for structural analysis, to find out if the frequency range of the analysis is sufficient to cover the modes of interest. Appendix A, Calculation of Effective Mass, of this Handbook provides a more general definition of effective and residual masses, and section 5.5 of this Handbook describes a method of calculating force limits based on this model, i.e., the Complex TDOF System Method.

Figure 6, Apparent Mass, Asymptotic Mass, Modal Mass, and Residual Mass of Longitudinally Vibrating Rod, Excited at One End and Free at Other End, shows the apparent, effective modal, and residual masses of a rod driven axially at one end (Scharton, 1997 [23]). Figure 6 also shows the asymptotic mass, which is the frequency average or non-resonant (Q=1) absolute magnitude of the apparent mass. (The asymptotic form of a frequency response curve is also sometimes called the skeleton (Neubert, 1987 [24]).)

The mass characteristics of the test item can be measured during a force limited vibration test and may be used to assess or update the calculated force limits. First, the magnitude of the drive point apparent mass, i.e., the ratio of total reaction force in the excitation direction to the input acceleration is measured during a low-level sine sweep, or random run. Then the apparent mass function is smoothed to determine the asymptotic mass, which must be a decreasing function of frequency, according to Foster's theorem [24]. The effective and residual masses of each mode may then be estimated from the asymptotic mass. Alternately, the modal effective mass for each distinguishable mode may be evaluated by equating the corresponding peak in the apparent mass curve to the sum of the residual mass and the product of the effective modal mass times the quality factor (Q), which may be determined from the half-power bandwidth. These mass characteristics of the test item may also be determined from an FEA of the test item driven at the base, with all the attachment points rigidly connected to a single node at the base, e.g., by using an RBE2 element in NASTRAN, and with a force link to determine the drive force.

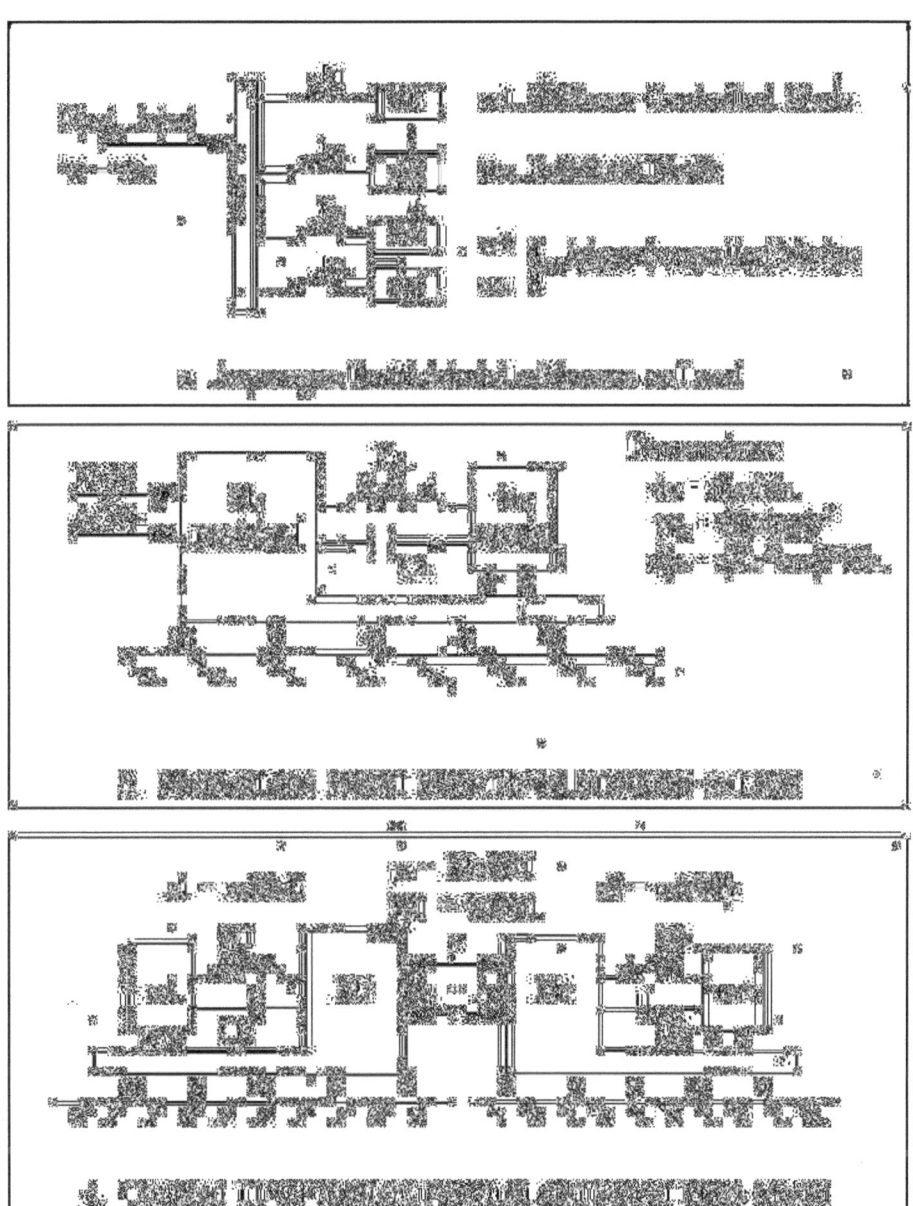

Figure 5—Complex TDOF System with Residual and Modal Masses

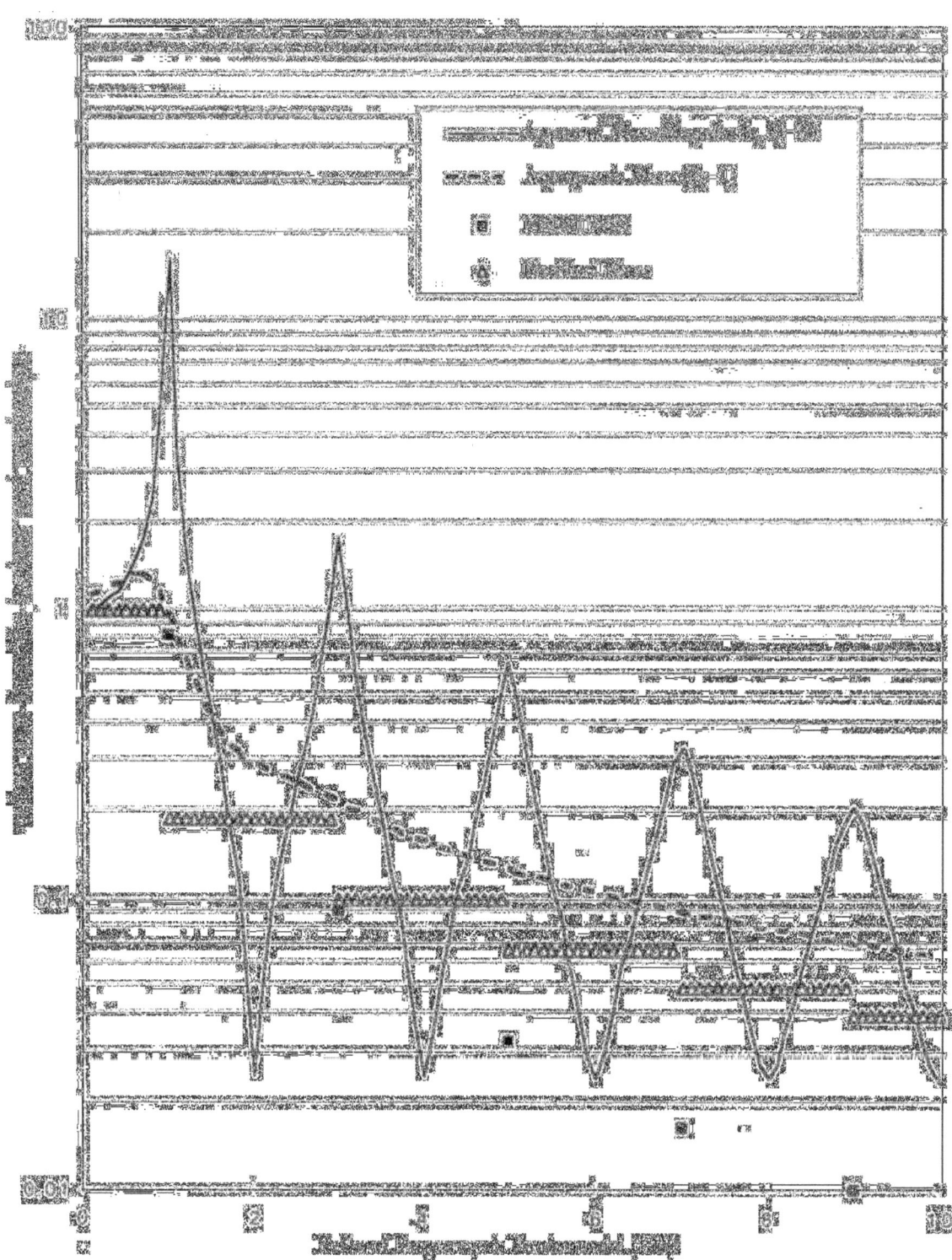

Figure 6—Apparent Mass, Asymptotic Mass, Modal Mass, and Residual Mass of Longitudinally Vibrating Rod, Excited at One End and Free at Other End

The apparent mass looking back into the flight mounting structure may also be determined with an FEA in a manner similar to that described for the test item. Alternately, the apparent mass of the mounting structure may be experimentally measured with a modal hammer incorporating a force gage. (The test item should not be attached to the mounting structure during the FEA or during experimental measurements.) The measurement process involves tapping at representative positions where the test item attaches to the flight mounting structure and computing the FRF of the ratio of the force to the acceleration, which may be measured with an accelerometer mounted temporarily on the structure near the hammer impact point. Some judgment is involved in combining the experimentally determined apparent masses measured at multiple attachment points to obtain a single-node model of the mounting structure apparent mass (Piersol, et al., 1989 [25]). At low frequencies and closely spaced test item attachment points, the apparent mass measured by tapping at any of the attachment points may yield the total mounting structure interface apparent mass; whereas, at high frequencies and widely spaced attachment points, the apparent masses measured by tapping at the different attachment points may represent only fractions of the total mounting structure interface apparent mass. Also, whether conducting an FEA or experimentally measuring the apparent mass of a mounting structure, it is important to decide how much of the adjacent structure needs to be represented. The criterion is to include just enough of the mounting structure so that the mounting structure effective modal and residual masses are accurately represented in the frequency range of the test item resonances on the shaker. (If too much of the mounting structure is included in the FEA, the upper frequency of the analysis may be too low.)

5.3 Semi-Empirical Force Limit Equation

A semi-empirical equation for relating the amplitude of the force limit (F) to the amplitude of the input acceleration specification (A) may be written in the form [2]:

$$F(f) = C\ M_o\ A(f), \qquad f < f_b \qquad \text{(Eq. 1a)}$$

$$F(f) = C\ M_o\ (f_b/f)^n\ A(f), \qquad f \geq f_b \qquad \text{(Eq. 1b)}$$

where:

C is a dimensionless constant (frequency independent), which depends on the configuration
M_o is the total mass of the test item
f is frequency
f_b is a break frequency
n is a positive constant.

The corresponding equation for random vibration tests is:

$$S_{FF}(f) = C^2\ M_o^2\ S_{AA}(f), \qquad f < f_b \qquad \text{(Eq. 2a)}$$

$$S_{FF}(f) = C^2\ M_o^2\ (f_b/f)^{2n}\ S_{AA}(f), \qquad f \geq f_b \qquad \text{(Eq. 2b)}$$

where:

S_{FF} is the force power spectral density (PSD)
S_{AA} is the input acceleration PSD.

Comparing Eq. 1a with the response of an SDOF system excited at its resonance frequency, where the base reaction force is Q times the mass of the SDOF system times the base acceleration, it may be seen that the constant C effectively replaces the amplification factor Q. In a random vibration test, for example, the semi-empirical force limit effectively "clips" the PSD of the reaction force at a value of C^2, rather than letting it have a peak value of Q^2. (For simplicity, the distinction between the total mass and the effective modal mass of the dominant mode and a residual mass term is ignored in this comparison.)

The constants f_b and n in Eqs. 1b and 2b account for the decrease of the asymptotic mass of the test item with frequency. (See section 5.2 and figure 6 of this Handbook.) Before the vibration test, the break frequency (f_b) in Eqs. 1b and 2b is usually specified as the resonance frequency of the fundamental mode of the test item on the shaker. Sometimes, the resonance frequency is defined before the testing by modal test data or by a FEM analysis or both. In any case, the assumed value of the break frequency should be verified, and if need be, adjusted, after a low-level, non-force limited run with the test item mounted with force gages on the shaker. (The non-force limited run has to be sufficiently low in level, so as not to risk damage to the test item.) After this initial low-level run, the break frequency should be adjusted if any one of the following is noted: the measured fundamental resonance frequency differs significantly from that assumed, the measured force spectrum peak at a higher order resonance frequency exceeds the peak at the fundamental, or the measured force spectrum exhibits a series of nearly equal resonance peaks. In any case, the break frequency should be taken as the frequency at which the asymptote of the measured force spectrum starts to roll off.

The exponent n in Eqs. 1b and 2b is often equal to unity, but this again should be verified with a low-level, non-force limited run with the test item mounted with force gages on the shaker. If the break frequency (f_b) is properly selected, as described in the last paragraph, the constant n will seldom be less than unity, but it may need to be greater to follow the rolloff of the asymptotic mass, to notch for a higher frequency resonance, or to notch for a fixture or shaker resonance. However, in no case, should the rolloff be so great as to result in a loss in the broadband or non-resonant, high-frequency response of the test item. And, the rolloff should not be so great as to inadvertently notch high-frequency test item resonances, where appropriate force limiting rationale has not been applied.

The force limit in the semi-empirical method is proportional to the acceleration specification at each frequency. Therefore, caution should be exercised in the application of the semi-empirical method to situations where the flight forces and accelerations peak at significantly different frequencies, e.g., in a situation where the force peaks at a launch vehicle mode and the acceleration peaks at a spacecraft mode or vice versa. (See, for example, the Gamma-Ray Large Area Space Telescope (GLAST) spacecraft flight case history in section 7.3 of this Handbook.) Finally, it is strongly recommended that the guidelines in section 6.0 of this Handbook, particularly Guideline 2 in section 6.2, be reviewed before using the semi-empirical force limiting equations.

5.4 Simple TDOF System Method of Deriving Force Limits

The simple TDOF system method of deriving force limits is described in [10]. A basic assumption behind the simple TDOF system method of specifying force limits is that there are vibration modes of the source in the same frequency regime as the vibration modes of the load and that these source and load modes are coupled. Another assumption is that the subject force is the result of the in-axis acceleration at the same interface, i.e., not the result of a cross-axis or a rotational motion. The basic model for the simple TDOF system method is shown in figure 7, Simple TDOF System of Coupled Oscillators. The model represents one vibration mode of the source (System 1) coupled with one vibration mode of the load (System 2). In the general case, the maximum interface force and the maximum interface acceleration occur at one of the two coupled-system resonance frequencies, but they need not occur at the same resonance frequency. However, the maximum response of the load, and therefore the maximum interface force, occurs for the special system where the uncoupled resonance frequency of the load equals that of the source; for this particular case, the maximum of the ratio of the interface force to interface acceleration occurs at the lower of the two coupled-system resonance frequencies [10]. For this case, excited by a white noise vibration source, figure 8, Normalized Force Specification from Simple TDOF System, shows the ratio of the interface force PSD (S_{FF}) to the interface acceleration PSD (S_{AA}), normalized by the load mass (M_2) squared, as a function of the ratio of load to source masses (M_2/M_1). The equations for calculating the curves shown in figure 8 are presented in Appendix B, Equations for Calculating the Simple TDOF System Force Limit, of this Handbook.

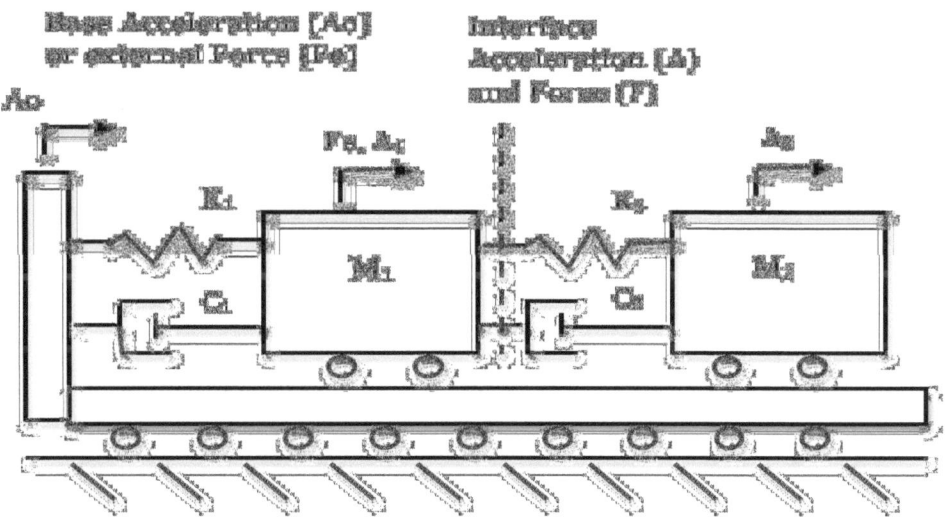

Figure 7—Simple TDOF System of Coupled Oscillators

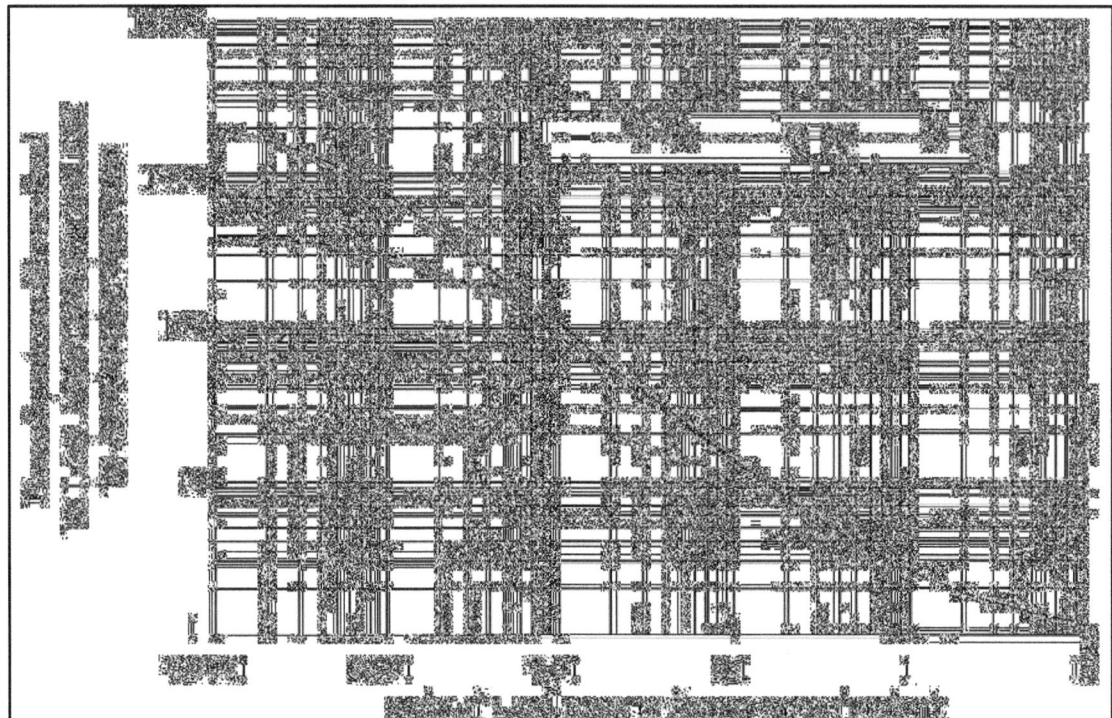

Figure 8—Normalized Force Specification from Simple TDOF System

When the ratio of load to source masses in figure 8 is very small, there is very little loading effect, and the asymptote of the ratio of force spectral density to the load mass (M_2) squared times the input acceleration spectral density is equal to the quality factor squared (Q^2). (The ratio of this asymptotic value of the force to the force limit at larger values of the load to source mass ratio (M_2/M_1) is sometimes called the knockdown factor.) The force limit itself is relatively insensitive to the damping at values of the load to source mass ratio (M_2/M_1) greater than 0.1. However, as the peak value of the unnotched force PSD is proportional to the quality factor squared (Q^2), the notch depth resulting from force limiting will be larger for lightly damped systems.

To use figure 8, the ratio of load to source masses (M_2/M_1) has to be determined. A limitation of the simple TDOF system model is that, as it does not take into account both the modal and residual masses, there is some ambiguity as to what masses to use in figure 8. For the simple TDOF method, it is recommended that the asymptotic apparent masses of the load and source be used to determine the mass ratio for the simple TDOF system, because the asymptotic apparent mass envelops both the modal and residual masses. (See the analysis of longitudinal vibrations in a rod in figure 6.) If both the residual and modal effective masses of the source and the load are available, from a FEM analysis or from tests, then it is recommended that the complex TDOF system model that is also discussed in [10] be used to determine the force limit.

The masses in the simple TDOF system model and the modal and residual masses in the complex TDOF system are, in general, functions of frequency. Usually, these masses need be evaluated only at the fundamental vibration mode of the test item on the shaker. However, if it is necessary or desired to evaluate the masses as a function of frequency, e.g., when notching of higher order modes is critical, it is suggested that a third-octave band analysis be utilized and that the results be averaged over the band.

Comparison of the ordinate in figure 8 with Eqs. 1 and 2 shows that, if the load mass (M_2) used to nondimensionalize the ordinate is taken as the asymptotic mass of the load, then the ordinate corresponds to the constant C^2 in the semi-empirical method. An advantage of the semi-empirical method and the simple and complex [10] TDOF system methods is that the force limit may be calculated using the input acceleration test specification as the reference acceleration, which, however, is not the case for the more detailed impedance method discussed in the section 5.6 of this Handbook.

5.5 Complex TDOF System Method of Deriving Force Limits

Appendix C, Tables for Complex TDOF System Method of Calculating Force Limits, of this Handbook tabulates the results of a force limiting analysis based on the complex TDOF system shown in figure 5. The derivation of this method and the governing equations are too complex to describe here, but they are delineated in [10]. This model is more complex than the simple TDOF system method described in the section 5.4 of this Handbook but offers the advantage that the masses involved are precisely defined as effective modal and residual masses; therefore, the method may be applied without the aforementioned ambiguity. The complex TDOF system method is particularly useful in conjunction with FEM analyses, where the effective modal masses are readily available. As with the simple TDOF system method, the normalized force spectra tabulated in Appendix C may be interpreted as C^2 in the semi-empirical method of section 5.3 of this Handbook.

5.6 Equivalent Circuit Impedance Method of Deriving of Force Limits

In the following discussion, frequency-dependent analytical expressions for the acceleration and force at the interface between a source and a load are derived using equivalent circuit theorems. The reference quantity in these expressions is the frequency-dependent free acceleration ($A_s(f)$) of the source, which can be calculated with some type of frequency response analysis, e.g., using a FEM to analyze the response of the unloaded source to external excitation, such as an acoustic or vibratory load. Alternately, if the free vibration of the source is determined from experimental measurements or from flight data, it is essential to utilize the raw spectral data from the same system used for the impedance analysis. Frequency envelopes of data or vibration specifications derived from envelopes are not appropriate inputs for the impedance method.

Using Norton's or Thevinen's equivalent circuit theorem, the ratio of the acceleration (A) acting at the interface of a coupled source (s) and load (l) to the free acceleration of the source (A_s) is given by:

$$A(f)/A_s(f) = M_s(f)/[M_s(f)+M_l(f)] \qquad \text{(Eq. 3)}$$

where:

M_s and M_l are the apparent masses of the source and the load, respectively (Neubert, 1987 [26]).

All the quantities in Eq. 3 are functions of frequency, and it is necessary to take the magnitude of the final result because, in general, the apparent masses are complex numbers. (The free acceleration of the source and the individual apparent masses of the source and load are defined at the interface junction, with the load and source disconnected. However, in the calculation of the apparent masses, it is essential that the senses of the applied forces and resultant accelerations at the interface satisfy geometric and force compatibility. One way of ensuring this is to calculate the sum of the source and load apparent masses (in the denominator of Eq. 3) with the source and load connected.)

As the load apparent mass (M_l) is equal to the ratio of the interface force (F) to the interface acceleration (A), Eq. 3 may be rewritten as:

$$F(f)/A_s(f) = M_s(f)*M_l(f)/[M_s(f)+M_l(f)] \qquad \text{(Eq. 4)}$$

Eq. 4 may be used to derive a force limit if the free acceleration of the source is known. It is important to re-emphasize that the free acceleration of the source in the Eqs. 3 and 4 is not to be confused with the acceleration test specification, which is typically a frequency envelope or empirical estimate of the source acceleration that lacks the detailed frequency information needed to implement the impedance method. If the acceleration test specification were erroneously used in place of the frequency dependent source acceleration in Eq. 4, the maximum value of the calculated interface force would be comparable to that which would occur in an unlimited, unnotched vibration test.

Figure 9, Impedance Method Results for TDOF System with Two Identical Oscillators (charts a through f), shows results calculated with the impedance method for an example consisting of two identical oscillators, connected in series. The parameters of the example, as defined in figure 7, are $A_o = 1$, $F_e = 0$, $M_1 = M_2 = 1$, $K_1 = K_2 = 1$, $C_1/M_1 = C_2/M_2 = 1/Q$, and $Q = 50$. The interface acceleration (A) in figure 9e, calculated from Eq. 3, has a maximum value of 51.76 at the coupled-system lower resonance frequency of 0.62, and the interface force (F) in figure 9f, calculated from Eq. 4, has a maximum value of 84.08 at the coupled-system lower resonance frequency.

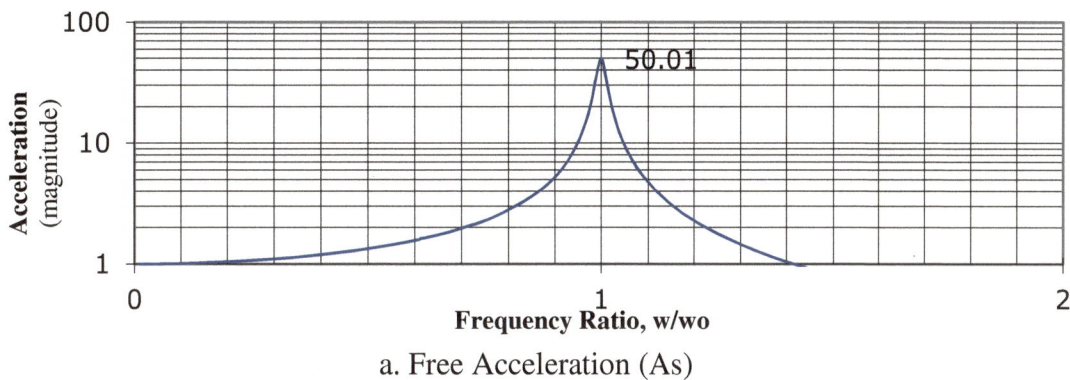

a. Free Acceleration (As)

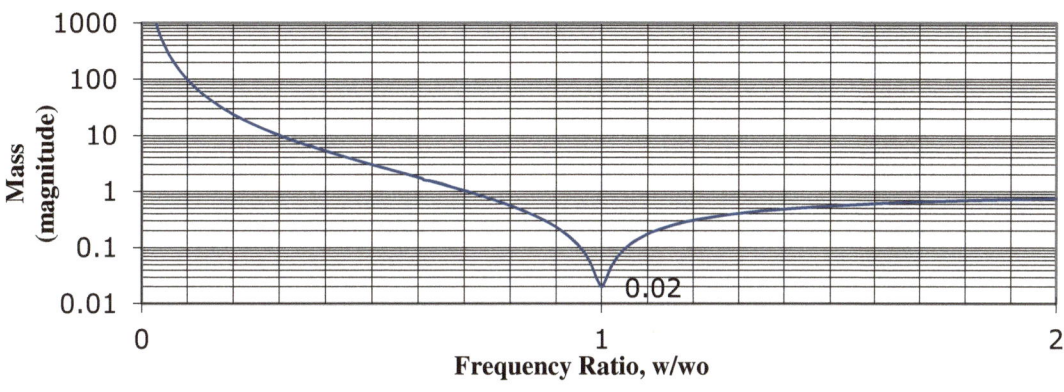

b. Source Apparent Mass (M_1)

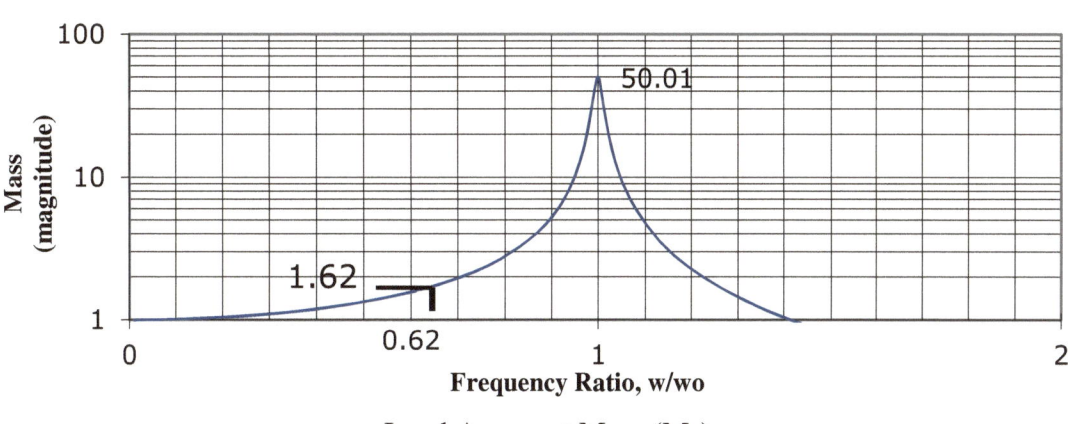

c. Load Apparent Mass (M_2)

Figure 9—Impedance Method Results for TDOF System with Two Identical Oscillators (1 of 2)

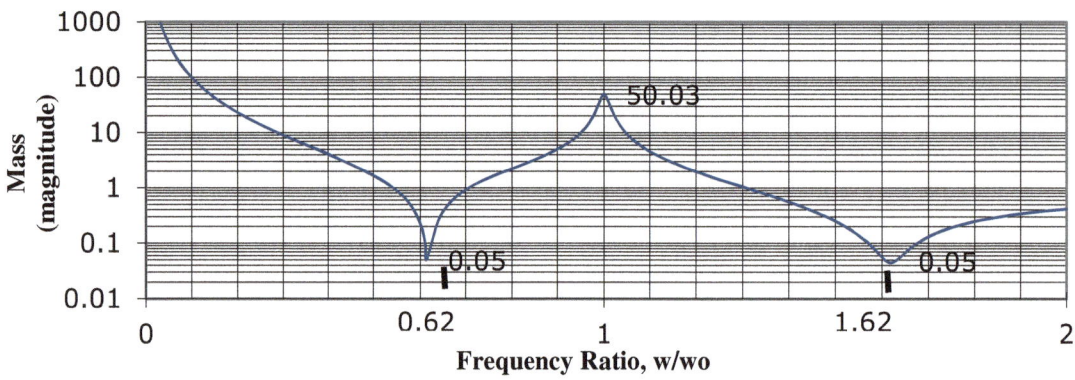

d. Sum of Source and Load Apparent Masses ($M_1 + M_2$)

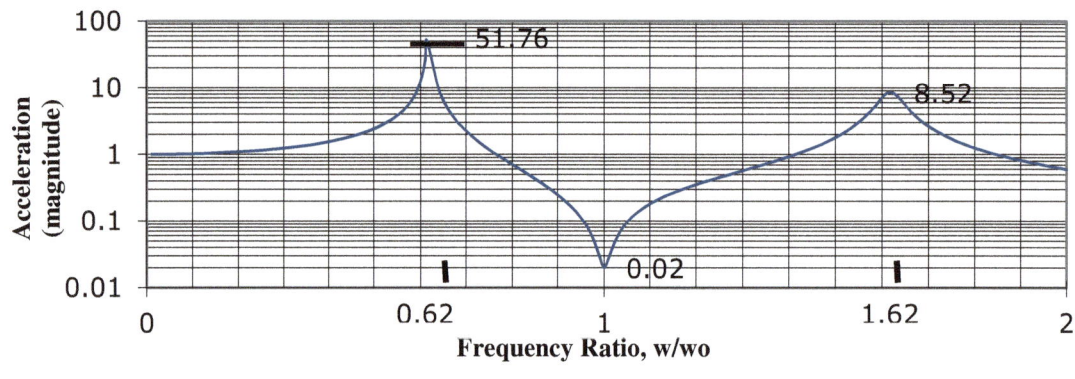

e. Interface Acceleration (A)

f. Interface Force (F)

Figure 9—Impedance Method Results for TDOF System with Two Identical Oscillators (2 of 2)

Note that, as shown in [10], the maximum value of the coupled-system interface force may be calculated by multiplying the value of the coupled-system interface acceleration, which may be taken as the acceleration test specification, by the load apparent mass, evaluated at the lower coupled-system resonance frequency, $M_2 (0.62) = 1.62$ (figure 9c), which is the square root of the ordinate value 2.64 in figure 8 for $M_2/M_1 = 1$. Another way of saying this is that the force in a vibration test should be limited to that calculated from the load apparent mass evaluated at the coupled-system resonance frequencies. This is preferred instead of letting the force go to the peak value of the load apparent mass, which occurs at the fixed-base resonance frequency of the load mounted on the shaker. (The unnotched interface force would be ~2500, if the maximum acceleration of figure 9e were used as the input acceleration at the fixed-base resonance frequency of the load mounted on the shaker.) The load apparent mass can be measured during a low-level vibration test with force gages, but the coupled-system resonance frequencies have to be calculated with a FEM.

The impedance method results presented herein match the values that would have been calculated by a coupled system analysis. Finally, it should be noted that application of the impedance approach is greatly simplified in cases where the interface connections can be reduced to a single node model, such as that considered herein.

5.7 Flight and Ground Test Scaling of C^2 in Semi-Empirical Method

To scale reference force data measured in a previous flight to derive force limits for a new vibration test, it is necessary to show similarity between the new configuration and the reference data configuration. As shown in section 5.4 of this Handbook, of particular importance is the ratio of the asymptotic mass of the test item (the load) to that of the flight mounting structure (the source) at the fixed-base resonance frequency of the load, i.e., the resonance frequency of the test item on a shaker.

Analytical and laboratory investigations have been conducted into the range of values of C in the semi-empirical force limit equation (Soucy, et al., 2005 [27]; Soucy, et al., 2005 [28]; Soucy, et al., 2006 [29]). In these studies, 134 configurations were investigated with the test item masses varied from 7.1 kg (15.7 lb) to 20.6 kg (45.4 lb) and its fundamental frequency varied from 97 Hz to 324 Hz. In all cases, the mounting structure had at least three modes with frequencies below the hard-mounted fundamental frequency of the test item. Table 1, Summary Table for C^2 Values in [29], shows the range of C^2 values measured for the different configurations tested; the majority of the values of C^2 were less than 5. However, it should be cautioned that the majority of the measured values of C^2 being less than 5 may have been associated with their particular design philosophy: "The final concept consists of the mounting structure being designed as a structure similar to the test item. However, it is much more flexible than the test item to be representative of typical real-life situations." [29] Again, it is recommended to take into consideration the ratio of the load to source asymptotic masses when determining a C^2 value for a new configuration.

Table 1—Summary Table for C^2 Values in [29]

C^2 Range	All Cases	Lateral Cases	Vertical Cases
$C^2 < 2$	43%	18%	62%
$C^2 < 5$	80%	66%	90%
$C^2 < 10$	92%	82%	100%
$C^2 < 20$	99%	97%	100%
$C^2 < 24$	100%	100%	100%

5.8 FEM Analysis of Force Limits

The most straightforward way to use a FEM to calculate force limits is to do a vibration analysis of the test item mounted on the flight mounting structure, which is in turn excited with a simulation of the flight environment, i.e., swept sine or random vibration, acoustic excitation, etc. Then, the force and acceleration at the test item and mounting structure interface are enveloped. The enveloping is necessary because the force limit specification should be relatively smooth, similar to acceleration specifications. Generally, the mounting structure is defined as the structure associated with the next level up in the assembly, e.g., the mounting structure on an instrument might be the spacecraft and that of the spacecraft might be the launch vehicle. Sometimes, it is practical to include only a portion of the mounting structure in the analysis, in which case the extent of the mounting structure included in the model should be big enough so that there are several modes below and in the vicinity of the fixed-base resonance frequency of the test item mounted on a shaker. However, the mounting structure model should not be so large that the credible frequency range of the FEM does not extend at least an octave or so above the fixed-base resonance frequency of the test item. Finally, it is important to realize that the biggest peak in the coupled-system interface force is likely to be at a frequency that is below the fixed-base resonance frequency of the test item. So, when developing the force specification, it is necessary to take the largest value of the coupled-system interface force to be the force limit at the fixed-base resonance frequency of the test item, where the largest forces will be generated when the test item is mounted on the shaker.

Four other ways of utilizing a FEM to develop force limits are (1) to calculate the asymptotic masses of the load and source to be used with the simple TDOF system described in section 5.4 of this Handbook, (2) to calculate modal and residual mass needed for the complex TDOF described in section 5.5 of this Handbook, (3) to calculate the free acceleration and the load and source apparent masses needed in the impedance method discussed in section 5.6 of this Handbook, and (4) to calculate the coupled-system resonance frequencies so that the measured (or calculated) load apparent mass can be evaluated at these frequencies, as suggested in section 5.6 of this Handbook.

One example (Staab, et al., 2012, [30]) of a FEM analysis of force limits involved the extension of the semi-empirical force limiting vibration method for use in analytical models. In this study, the steps in performing a force limiting analysis are reviewed, and the analytical results are compared to test data recovered during the Communications, Navigation, and Networking reConfigurable Testbed (CoNNeCT) force limiting random vibration qualification test conducted at NASA Glenn Research Center (GRC). A compilation of lessons learned and considerations for future force limiting tests are also included.

5.9 Quasi-Static Load Limits

The design of aerospace components with low-frequency primary resonances (below 50 or perhaps 80 Hz) is usually based on the quasi-static limit load (QSLL), which may be initially derived from a database, e.g., the launch vehicle user's guide, or from a curve, e.g., a mass acceleration curve, or, later in the program, from a coupled launch vehicle and spacecraft loads analysis (CLA). The QSLL is typically used to define the load or force at the base of the component, but it is usually expressed as an acceleration acting at the center of gravity (CG) of the system, as Newton's second law says that the external force on a system of masses equals the total mass times the acceleration of the CG.

Sometimes, it is necessary to achieve the QSLL acceleration or the QSLL times a test margin in a vibration test (usually in a sine dwell or sine burst test), and it is almost always necessary that the vibration response in any other vibration test should not exceed the QSLL or, again, the QSLL times some margin. However, in a vibration test, the CG acceleration may be difficult to measure accurately with an accelerometer, particularly at higher frequencies, because the CG may not even be located on the physical structure; furthermore, the CG of a flexible body is different for each mode shape (Vujcich and Scharton, 1999 [31]). Therefore, limiting the measured external force divided by the total mass of the test item is an alternative and often more accurate method of limiting the CG acceleration in a vibration test.

In a random vibration test, with or without force limiting, the RMS value of a response is defined as the square root of the integral of the response PSD over all frequencies. For design and test, it is common practice to use a peak factor of 3 to relate the RMS to the peak value of a response. However, a number of experimental investigations (Scharton, et al., 2006 [32]; Kolaini and Doty, 2007 [33]) have shown that peak factors of 4.5 to 5 can occur in random vibration and acoustic tests of aerospace hardware, so it is highly recommended that the higher values of peak factor be utilized, particularly for brittle structures. The peak factor can be determined during a test by recording the time history and by dividing the highest peak observed on the record by the RMS of the record. Then, it is prudent to choose a somewhat higher peak factor than that observed to reduce the probability that a higher peak level will occur in a subsequent test run.

Before running a test without force limiting or any other form of response limiting, the RMS acceleration response (σ_{CG}) of the CG may be estimated from Miles' equation (Miles, 1954 [34]):

$$\sigma_{CG} = [(\pi/2) \, Q \, f_o \, S_{AA}(f_o)]^{1/2} \qquad \text{(Eq. 5)}$$

where:

f_o is the fundamental resonance frequency of the test item
Q is the quality factor of the resonance
$S_{AA}(f_o)$ is the PSD of the acceleration input at the resonance frequency.

(Eq. 5 is an approximation for two reasons: it assumes a SDOF system response, and it ignores the mode shape in the evaluation of both the modal force and the modal response.)

If a force limited random vibration test is to be conducted using the semi-empirical force limit prediction method described in section 5.3 of this Handbook, then (Chang, 2002 [35]) has shown that the RMS acceleration response (σ'_{CG}) of the CG may be estimated from:

$$\sigma'_{CG} = [2 \, C \, f_o \, S_{AA}(f_o)]^{1/2} \qquad \text{(Eq. 6)}$$

where:

the quality factor Q in Eq. 5 has been replaced by $(4/\pi)*C$, where C is the constant in the semi-empirical method.

(Eq. 6 is an approximation for the same reasons as Eq. 5.)

6. THREE GUIDELINES

6.1 Guideline 1: Use Force Limiting Only for Highly Resonant Test Articles

Use force limiting for structural-like and other highly resonant test articles, e.g., antennae, optics, mirrors, cryogenics, large instruments, and spacecraft. Do not use force limiting for non-resonant test articles, e.g., hard-mounted electronic boxes or heavily damped test items. A rule of thumb is not to use force limiting for items with Q < 2, i.e., those with a resonance peak less than 6 dB. From figure 8, it may be seen that a value of C < 2 corresponds to a value of $M_2/M_1 > 0.5$, so unless there are analyses or data to confirm this relatively large mass ratio, force limiting should not be used if Q < 2. (As discussed in the second guideline below, the mass ratio should always be a consideration when force limiting is employed.) Sometimes, it is necessary to reduce the response of test items when the amplification is less than Q < 2, e.g., to reduce the response to the flight limit loads, to avoid overtesting a particular piece of equipment or to protect the shaker or fixturing. In these cases, one should consider reducing the acceleration input rather than notching at the resonance frequency. Unfortunately in these situations, force limiting is sometimes used in lieu of trying to change the acceleration specification, which is often difficult or impossible to change because of programmatic issues or the complexities involved in the flow down of the requirements.

6.2 Guideline 2: Use Appropriate Rationale for Deriving Force Limits

The primary purpose of the semi-empirical method (section 5.3 of this Handbook) is to provide a simple framework for comparing flight force data and the results of various other force prediction methods. Therefore, one should not use the semi-empirical method by itself with an arbitrarily chosen C^2, with a C^2 based solely on the test item mass, or with a C^2 value selected so as to reduce the response to a predetermined level. Do justify the choice of C^2 with rationale based on consideration of the flight mounting configuration of the test item. Some examples of appropriate considerations for selecting C^2 include: the simple and complex TDOF system methods presented in sections 5.4 and 5.5 of this Handbook, respectively; the equivalent circuit method discussed in section 5.6 of this Handbook; a FEM analysis of the flight mounting configuration addressed in section 5.8 of this Handbook; and reference to the C^2 values measured in flight or ground tests for similar configurations discussed in section 5.7 of this Handbook. Also, it is a good idea to survey the literature to determine the history of C^2 values used for similar test items by organizations with extensive experience in applying force limiting. In addition, be cautious using the semi-empirical method or the simple TDOF method when the maximum flight forces and accelerations occur at widely different frequencies. (See section 7.3 of this Handbook.)

6.3 Guideline 3: Avoid Excessive Notching

Excess notching should be avoided. As the first guideline implies, there should be at least a 6-dB resonance remaining after notching. Also, it is recommended that one seek the counsel of colleagues and/or a supervisor if the calculated notch depth or that measured during low-level tests exceeds 14 dB. In the evaluation of the notches, it should be kept in mind that lightly damped resonances result in deeper notches. A simple guideline for accessing notch depth for lightly damped resonances is to compare the notch with Sweitzer's notching criterion (Sweitzer, 1987 [36]) discussed in Appendix D, Sweitzer's Notching Criterion, of this Handbook, i.e., that the input acceleration is notched so as to reduce the response to half its original value in decibels. (For example, a 20-dB resonance becomes a 10-dB resonance and a 10-dB notch.)

Excessive notching of the input acceleration often results when it is desired to reduce the RMS response at some position on the test item by a large amount, say 50 percent or more. Figure 4 shows that, for an SDOF system, 14 dB of notching is required to reduce the RMS response by 50 percent, and no amount of notching will reduce the RMS response by more than about 68 percent. As multiple modes contribute to the response measured at a single position, notching of a single mode has even less effect than the aforementioned numbers. By contrast, for a linear system, the RMS response at every measurement position will be reduced by 50 percent if the input acceleration specification is reduced (at all frequencies) by 6 dB. As this comparison illustrates, do take into consideration both the force limit and the input acceleration specification when it is desired to reduce the severity of a vibration test, and do not use force limiting to compensate for an acceleration specification that may be too high.

7. FLIGHT AND GROUND VIBRATION TEST DATA

Three experiments in which vibration forces were measured are described herein. In two cases, the flight measurements of vibration force and acceleration are compared with the corresponding data from the ground vibration tests of the payloads.

7.1 Shuttle Vibration Force (SVF-2) Experiment on Space Shuttle Space Transportation System-96 (STS-96)

Figure 10, SVF-2 Experiment on STS-96, is a photograph of the SVF-2, which flew in May of 1999 [15]. The SVF experiment also flew on Shuttle flights STS-90 and STS-102, but no force data were obtained on those two flights. The SVF-2 experiment utilized a Hitchhiker (HH) canister attached to the Shuttle sidewall via an adapter beam. The adapter beam also held a second HH experiment; the SVF-2 is the HH canister on the right in figure 10. Four triaxial force gages were located between the SVF-2 canister and the adapter beam, and two triaxial accelerometers (along with the signal processing and recorders) were located inside of the canister, as shown in figure 11, Hitchhiker Canister for SVF-2. However, the accelerometer at the canister CG, which is the lower of the two accelerometers indicated in figure 11, did not provide good data on the SVF-2 flight.

Figure 10—SVF-2 Experiment on STS-96

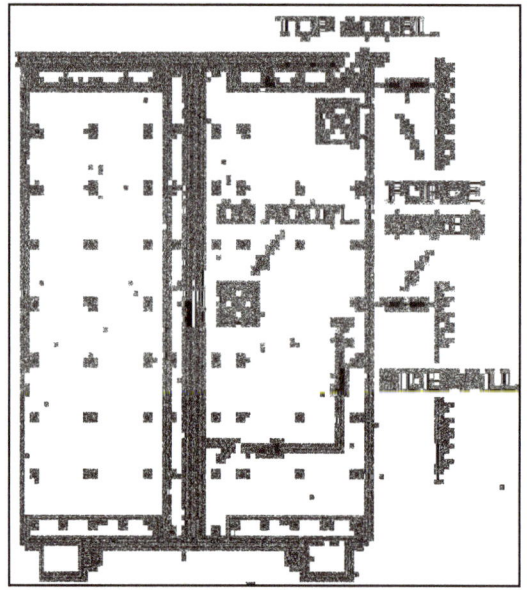

Figure 11—Hitchhiker Canister for SVF-2

For brevity, only the acceleration and force data measured in the Y-axis, normal to the Shuttle sidewall, are discussed. The Y-axis random vibration is generally larger than that parallel to the sidewall, because acoustic excitation is the primary source of random vibration of the sidewall.

Figure 12, Force Limit in Vibration Test of SVF-2 Canister, and figure 13, Acceleration Input in Vibration Test of SVF-2 Canister, show the PSD of the force and acceleration measured in the SVF-2 canister force limited vibration qualification test, which was conducted approximately 2 years before the flight. The force limit of 10,000 lb^2/Hz for the vibration test was derived using the semi-empirical method of Eq. 1b with an input acceleration spectrum S_{AA} of 0.04 G^2/Hz (figure 13), a test item weight M_o of 104 kg (230 lb), and a C^2 of 4. This resulted in a 16-dB notch at the fundamental resonance of approximately 150 Hz. (The calculated force limit of 8,464 lb^2/Hz was rounded off to 10,000 lb^2/Hz for the test.).

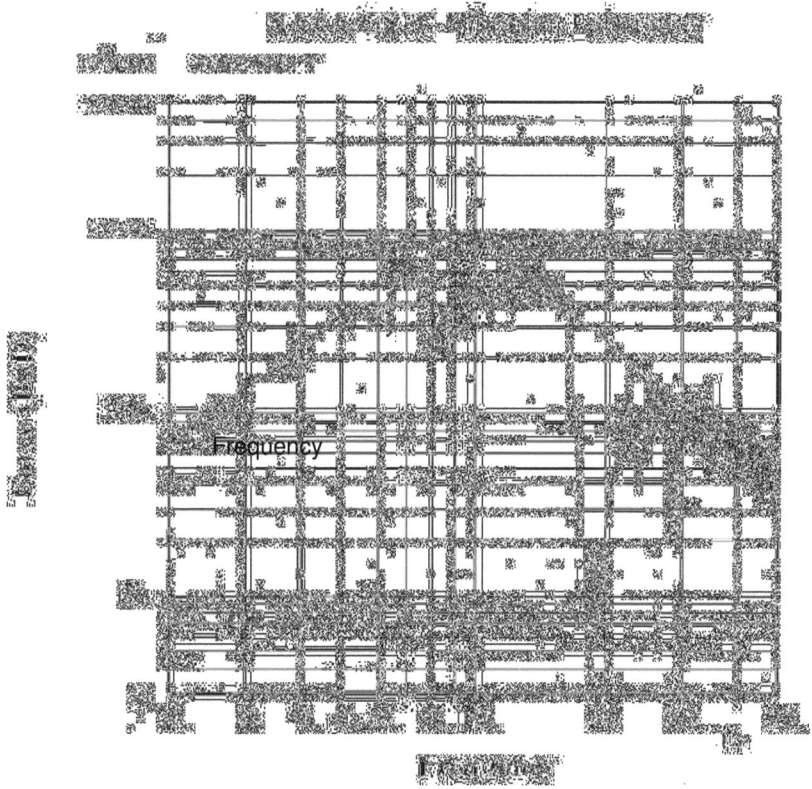

Figure 12—Force Limit in Vibration Test of SVF-2 Canister

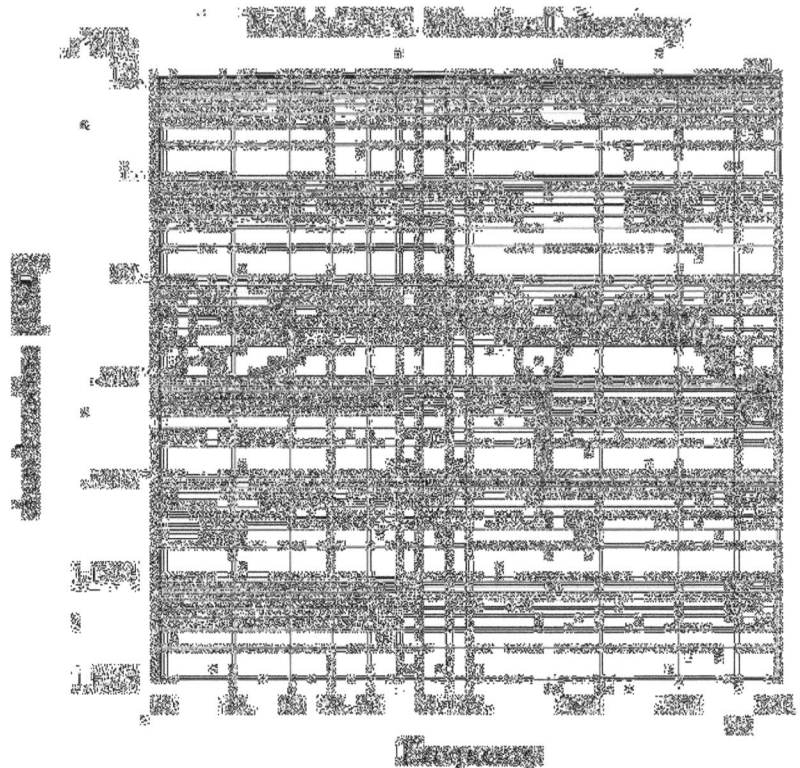

Figure 13—Acceleration Input in Vibration Test of SVF-2 Canister

The flight data shown in figure 14, Y-Axis Acceleration at Top of SVF-2, and figure 15, Total Y-Axis Force in SVF-2 Flight, are PSDs calculated during the time interval 7 < T < 9.5 seconds after ignition of the Shuttle main engine. The maximum acoustic and random vibration levels occurred during this time interval. The spectral analyses were conducted using MATLAB with an analysis bandwidth of 5 Hz.

Figure 14 shows the Y-axis acceleration measured in flight by the top accelerometer on SVF-2. The peaks in the flight acceleration spectrum of approximately 0.02 G^2/Hz are a factor of two below the 0.04 G^2/Hz (figure 13) acceleration input specification for vibration qualification tests of SVF-2 canisters. This is compatible with the NASA standard 3-dB margin in NASA-STD-7001, Payload Vibroacoustic Test Criteria. Also, it should be recognized that the specification is for the adapter beam input to the canister, whereas the measured data are actually responses of the canister at a position relatively close to the adapter beam attachment. Measurements directly on sidewall-mounted adapter beams for previous Shuttle flights indicate that 0.01 G^2/Hz is a typical value for the envelope of the input acceleration PSD (Talapatra, et al., 1983 [37]). The flight response measurement shown in figure 14 is consistent with the thesis put forth in [2] that there is little amplification between the vibration input and response in actual in-service configurations.

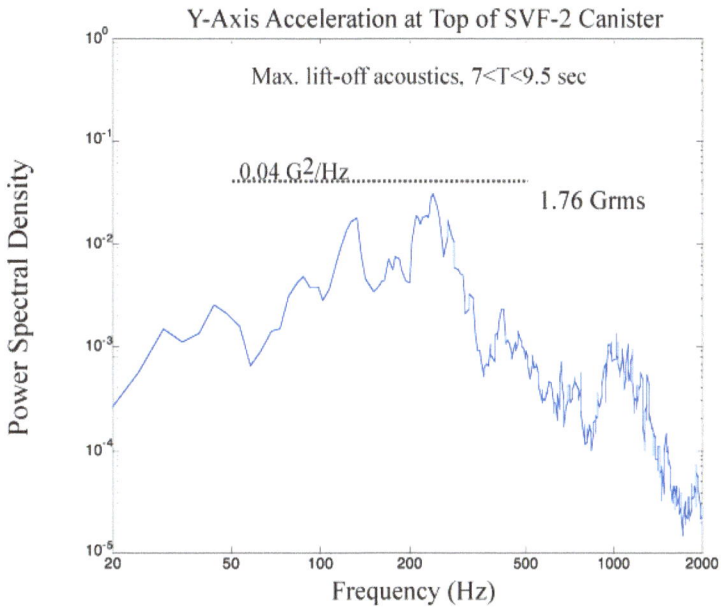

Figure 14—Y-Axis Acceleration at Top of SVF-2

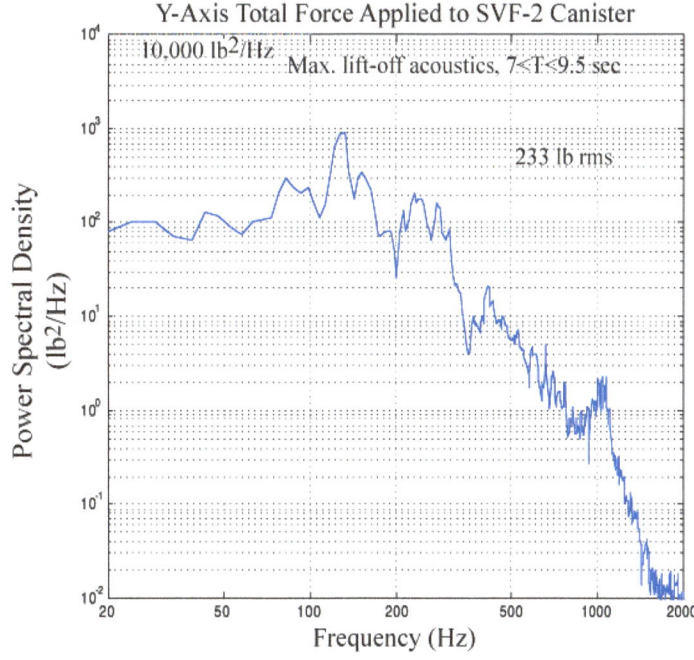

Figure 15—Total Y-Axis Force in SVF-2 Flight

Figure 15 shows the total Y-axis force measured in flight between the sidewall and the canister. The total force was obtained by summing in real time the Y-axis outputs of the four force gages located between the adapter beam and the canister. The measured PSD of flight force rolls off approximately 9 dB/octave at frequencies above 130 Hz, where it has a maximum value. The maximum force PSD of 1,000 lb^2/Hz measured in flight is an order of magnitude less that the 10,000 lb^2/Hz (figure 12) PSD force limit used in the vibration test of the SVF-2 canister. Even with this relatively high force limit, a notch of 16 dB (figure 13) resulted at the fundamental resonance frequency of the canister in the force limited vibration test.

If the flight input acceleration is assumed to have a maximum value of the acceleration PSD of 0.01 G^2/Hz, which is consistent with previous measurements of Shuttle sidewall vibration reported in [37] and with the response measurements in figure 14, then Eq. 2b with the measured maximum value of the force PSD of 1,000 lb^2/Hz (figure 15) and a canister weight of 104 kg (230 lb) yields a value of C^2 of 1.9 for the SVF-2 flight data.

7.2 Cosmic Ray Isotope Spectrometer (CRIS) Instrument on Advanced Composition Explorer (ACE) Spacecraft

The flight data described in [14] were measured at the interface of the CRIS instrument and the ACE spacecraft. The data were recorded during a 1-second interval corresponding to the time of maximum acoustic loading during the liftoff of the Delta II 7920-8 launch vehicle. Figure 16, CRIS Instrument on ACE Spacecraft Bus, shows the 30-kg (65-lb) CRIS instrument mounted on the left side of the ACE spacecraft bus, which is a two-deck octagon honeycomb structure, 1.6 m (65 in) across and 1 m (40 in) high.

Figure 16—CRIS Instrument on ACE Spacecraft Bus

Figure 17, Total Vertical Force in CRIS Random Vibration Test, shows the total vertical force in the CRIS random vibration test, and figure 18, Notched Acceleration Input in CRIS Random Vibration Test, shows the notched acceleration input in the test. A low-level sine-sweep vibration test of the CRIS instrument mounted on the 12 force gages indicated that the gages read only about 83 percent of the total weight of the instrument, so the force gage PSD measurements in the vibration test, as well as in flight, were increased by a factor of $(1/0.83)^2$ or 1.44. (The other 17 percent of the force goes through the force gage bolts.) Multiplying the 800 lb^2/Hz force limit in figure 17 by 1.44 and dividing by the square of the instrument weight of 30 kg (65 lb) and by the value of the acceleration specification of 0.12 G^2/Hz at 200 Hz yields a value for C^2 in Eq. 1b of approximately 2.3 for the vibration test force limit. A notch of 7 dB (figure 18) resulted at the fundamental resonance frequency of the instrument in the force limited vibration test.

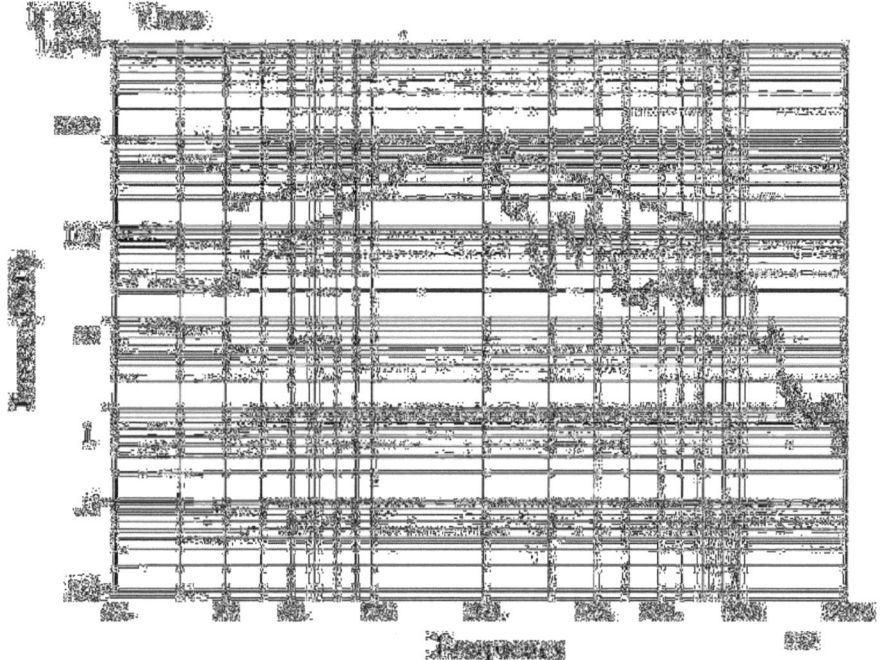

Figure 17—Total Vertical Force in CRIS Random Vibration Test

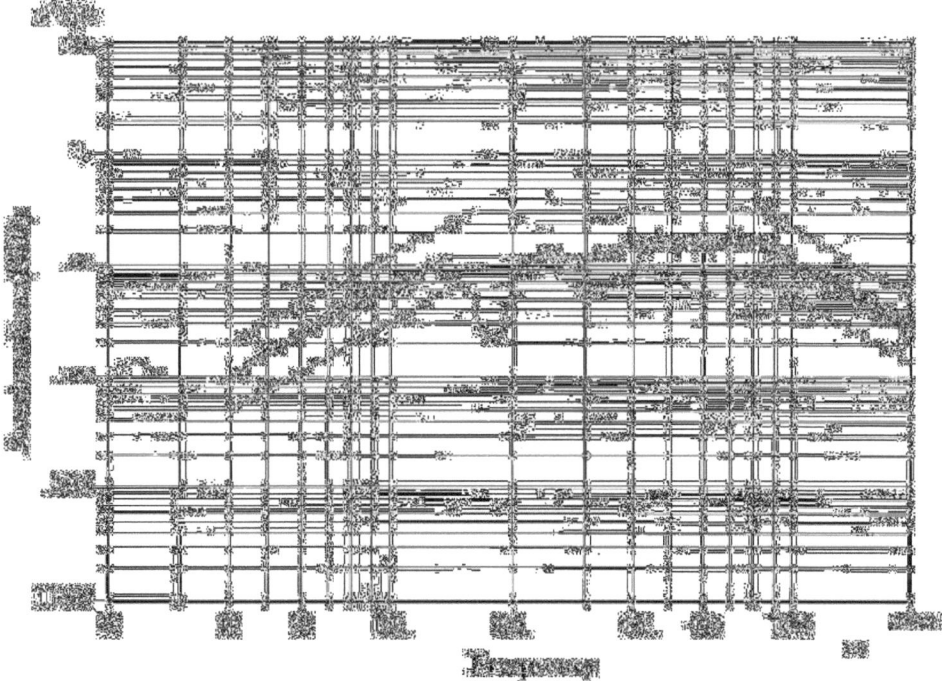

Figure 18—Notched Acceleration Input in CRIS Random Vibration Test

Figure 19, Spectral Density of Flight Normal Acceleration Measured near One Mounting Foot of CRIS Instrument, shows the PSD of flight normal acceleration measured near one mounting foot of the CRIS instrument, and figure 20, Spectral Density of Flight Normal Force Measured under CRIS Instrument, shows the PSD of flight total normal force measured under the CRIS instrument. Both the flight force and acceleration PSDs peak at the coupled-system resonance frequencies, approximately 33 Hz and 135 Hz. The flight force PSD decreases with frequency, above the 135-Hz resonance, where it has a maximum value, but the flight acceleration remains high above this frequency. (That both these coupled-system resonances occur at lower frequencies than the fundamental vertical resonance on the shaker, approximately 180 Hz, may be associated with rocking motion in the flight mounting configuration.)

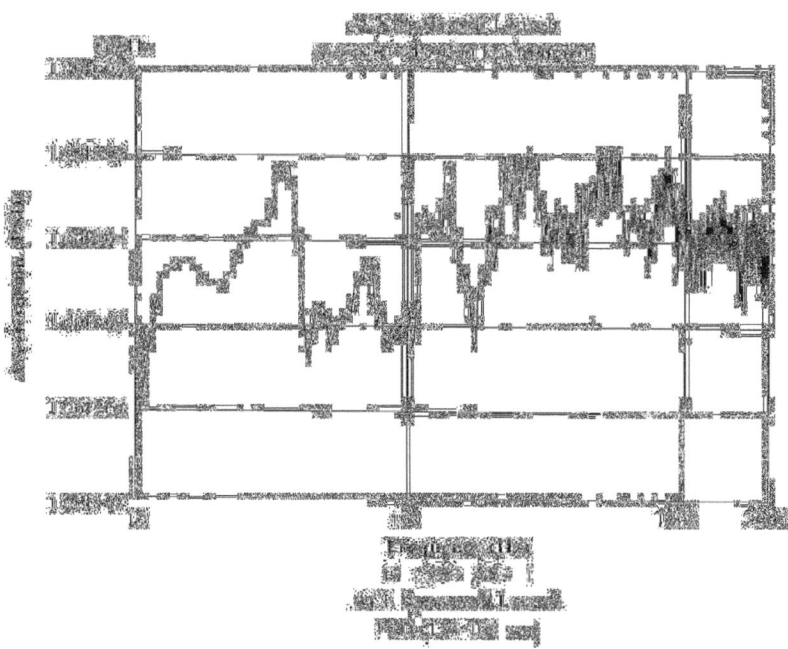

Figure 19—Spectral Density of Flight Normal Acceleration Measured near One Mounting Foot of CRIS Instrument

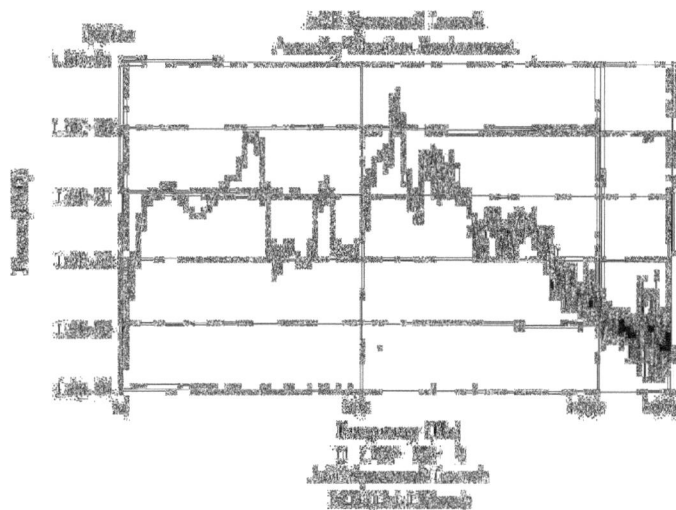

Figure 20—Spectral Density of Flight Normal Force Measured under CRIS Instrument

The maximum flight acceleration PSD of 0.001 G^2/Hz in figure 19 is two orders of magnitude lower than the acceleration PSD specification in the instrument random vibration test in figure 18, and the maximum force PSD in figure 20 is also approximately two orders of magnitude below the vibration test force limit in figure 17.

The ratio of the measured flight force and acceleration PSDs at the 135-Hz resonance frequency, where the force is a maximum, is approximately 5,000 lb^2. Applying Eq. 2a with the 1.44 force measurement correction factor and the instrument weight of 30 kg (65 lb) yields a value of C^2 of 1.7 for the CRIS flight data.

7.3 GLAST Spacecraft on Delta II Launch Vehicle

The Flight Force Measurement (FFM) project (Gordon and Kaufman, 2009 [38]) involved flight measurements of the accelerations and forces at the interface of the GLAST 4,378-kg (9,631-lb) spacecraft and the Delta II-6915 launch vehicle, which was launched in June 2008. To measure the forces, the Payload Adapter Fitting (PAF) was instrumented with 64 strain gages, which were calibrated with a static pull test during a ground vibration test. The objective of the project was to address two questions: Are acceleration measurements alone sufficient for flight correlation and reconstruction, and how much can the loads and, therefore, design/qualification requirements be reduced by also having force measurements?

The flight force and acceleration data measured during the FFM program are presented here as an example of the relationship between the flight forces and accelerations at a launch vehicle and spacecraft interface and particularly to illustrate that the maximum values of these forces and accelerations may occur at widely different frequencies, which violates fundamental assumptions in the semi-empirical and TDOF methods of deriving force limits. A second, somewhat surprising, result is that the amplification factor (Q) measured in flight at the first vertical resonance of the GLAST spacecraft at 31 Hz was similar to the Q measured in the vibration test, i.e., approximately a Q of 10 [38]. The GLAST data reported herein are not intended for use to predict the loads on other spacecraft.

Figure 21, GLAST Spacecraft Sine Vibration Test, shows the GLAST spacecraft axial sine vibration test conducted in September 2007. (The spacecraft was vibration tested in all three axes.) Piezoelectric force gages were installed under the PAF to validate the accuracy of the PAF strain gages, which would measure the interface forces in flight. The GLAST vibration test input levels were derived from the CLA liftoff event. The GLAST vibration test, however, was not force limited.

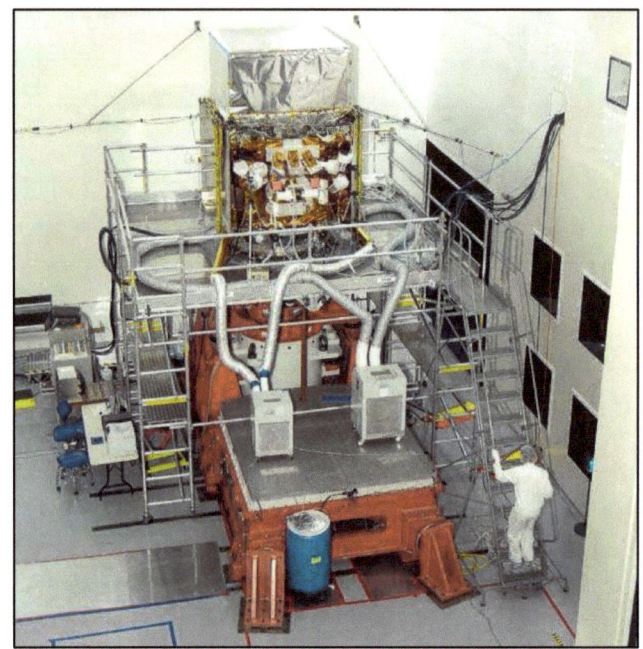

Figure 21—GLAST Spacecraft Sine Vibration Test

Figure 22, GLAST Spacecraft X-Axis (Lateral) Interface Flight Accelerations, and figure 23, GLAST Spacecraft X-Axis (Lateral) Base Flight Forces, show the X-lateral axis acceleration measured at the base of the PAF and the force measured by the PAF strain gages during the liftoff and aero-loads portions of the launch; figure 24, GLAST Spacecraft Z-Axis (Thrust) Interface Flight Accelerations, and figure 25, GLAST Spacecraft Z-Axis (Thrust) Flight Base Forces, show similarly the accelerations and forces in the Z-axial (thrust) direction (Gordon, et al., 2009 [39]). The accelerations are the average of six triaxial accelerometers mounted under the launch vehicle adapter (LVA), and the forces are the total forces measured in three axes with the LVA strain gage system.

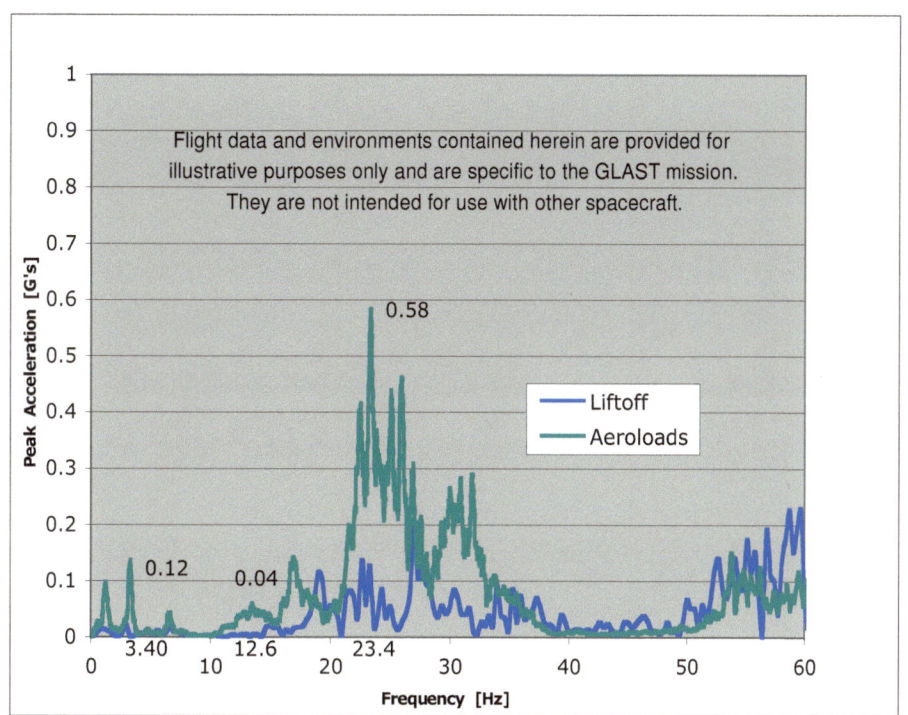

Figure 22—GLAST Spacecraft X-Axis (Lateral) Interface Flight Accelerations

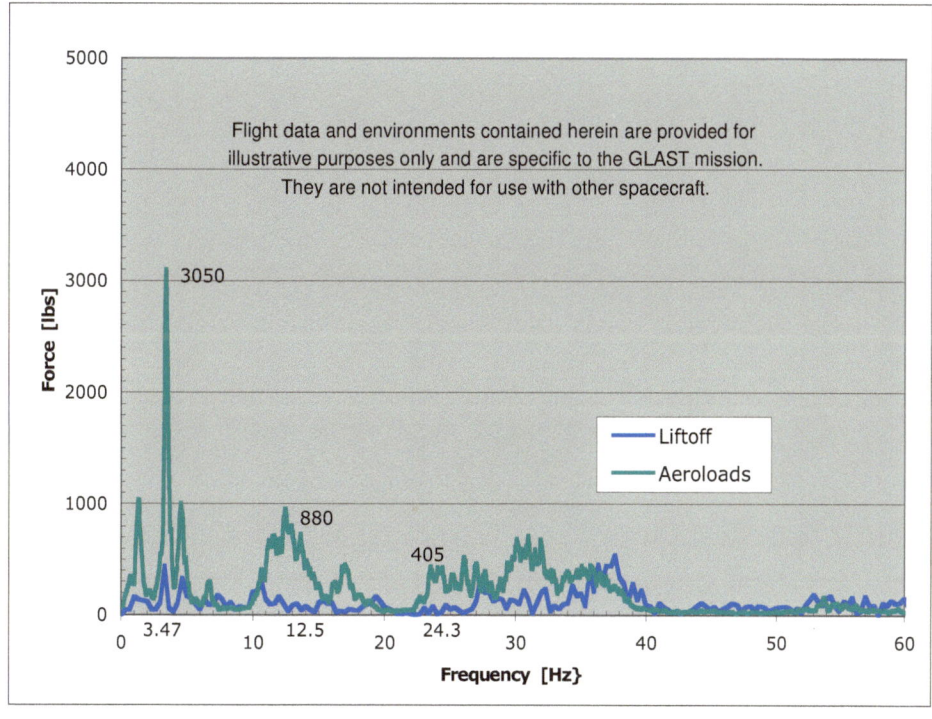

Figure 23—GLAST Spacecraft X-Axis (Lateral) Base Flight Forces

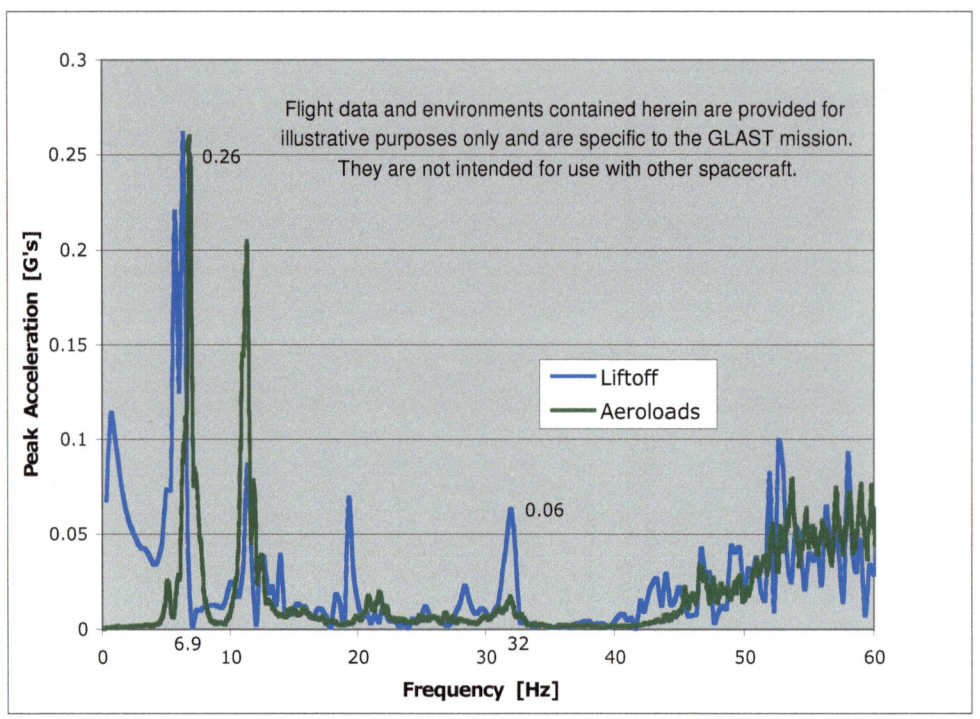

Figure 24—GLAST Spacecraft Z-Axis (Thrust) Interface Flight Accelerations

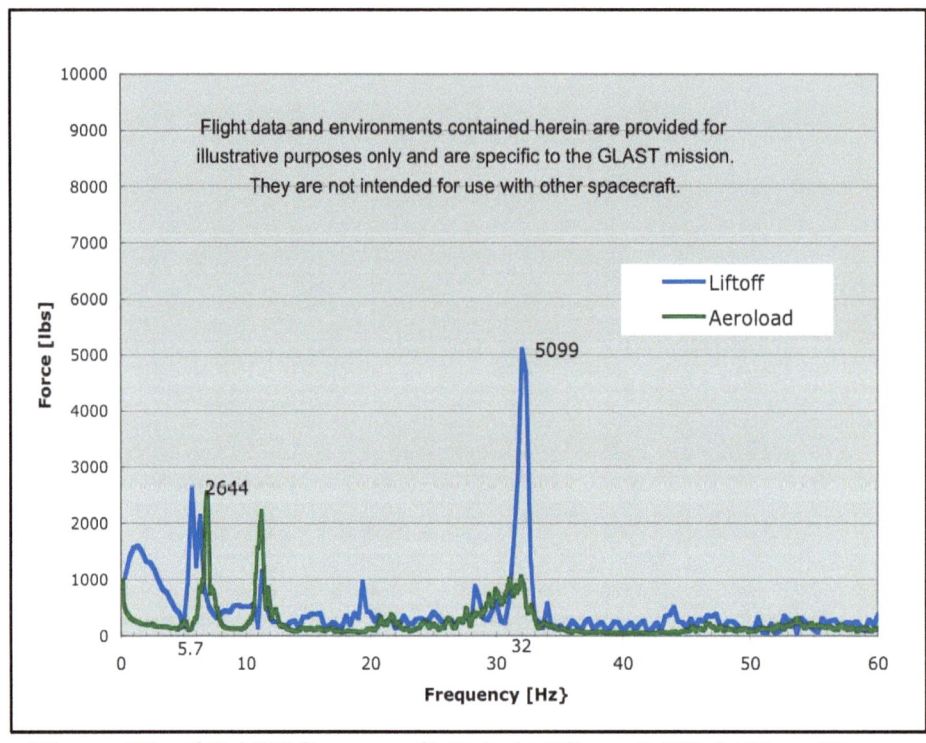

Figure 25—GLAST Spacecraft Z-Axis (Thrust) Flight Base Forces

The accelerations and forces during the aero-load portion of the launch were predominantly random vibration, but the accelerations and forces measured during liftoff involved transient, sine, and random vibrations. For comparison with sine vibration test results, it is traditional to analyze flight acceleration data using shock response spectra (SRS). However, SRS are not applicable to the analysis of base force data; therefore, an alternative random equivalent analysis method, based on the PSD calculated from the fast-Fourier transform (FFT) of the data, was utilized to calculate the flight acceleration and force data shown in figures 22-25. The random equivalent equates the amplitude of a sine input to that of a resonance peak in a random input, i.e.,

$$\text{Random equivalent} = PF * [(\pi/2) * f * PSD/Q]^{0.5} \quad \text{(Eq. 7)}$$

where:

PF is the peak factor (taken here as 5)
f is the analysis frequency
PSD is the power spectral density of the signal being analyzed
Q is the quality factor used in the analysis (here taken as 20).

(The factor in brackets in Eq. 7 differs from that in Mile's equation, Eq. 5, [34] by the Q^2, because Eq. 7 represents an input, not a response.)

As the flight accelerations and forces in figures 22 - 25 peak at different frequencies, some of which are associated primarily with the launch vehicle and some of which are associated primarily with the spacecraft, it is not possible to use these data to define the C in the semi-empirical method. However, it is interesting to estimate the amplification Q of the first axial mode of the spacecraft at 32 Hz from the flight data in figures 24 and 25: Q = 5,099 lb/[9,631 lb * 0.06 g] = 9, when the mode shape is neglected. So, to not exceed the 5,099-lb maximum axial force on the spacecraft at liftoff (figure 25), the acceleration input in an axial sine vibration test would have to be equal to or less than 0.06 G at 32 Hz (figure 24). If the acceleration input in the test were larger, the axial force would have had to be limited to the 5,099 lb. The axial sine vibration test acceleration input at 32 Hz specified in the Delta II Users Guide (06H0214, 2006 [40]) is 0.4 G plus a 3-dB test margin, but the input in the GLAST vibration test was limited to 0.15 G, based on a FEM analysis (NESC-RP-06-071, 2009 [41]).

Figure 26, GLAST Spacecraft Thrust Z-Axis Apparent Mass Measured in Ground Vibration Test and in Three Flight Regimes, shows the thrust axis apparent masses of the GLAST spacecraft calculated from the ratios of force to acceleration measured in the three different flight regimes (liftoff, aero-loading, and Main Engine Cut Off (MECO)) and compares these flight measurements with the apparent mass measured in the Z-axis sine vibration test. Figure 27, GLAST Spacecraft Lateral X-Axis Apparent Mass Measured in Ground Vibration Test and in Two Flight Regimes, is a similar comparison for the X-lateral axis. The agreement between the flight and ground test data is better in the thrust axis than in the lateral axis, where there appears to be a frequency shift. The comparisons between the flight and ground apparent mass measurements indicate the accuracy with which the forces in future spacecraft flights could be calculated from only flight acceleration measurements, combined with apparent mass measurements in ground vibration tests with the spacecraft mounted on force gages.

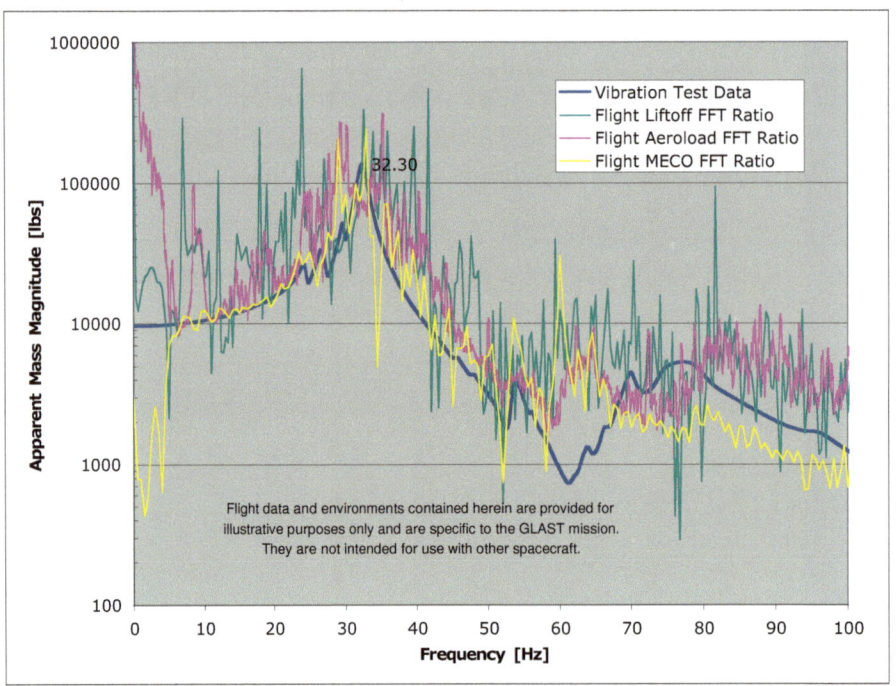

Figure 26—GLAST Spacecraft Thrust Z-Axis Apparent Mass Measured in Ground Vibration Test and in Three Flight Regimes

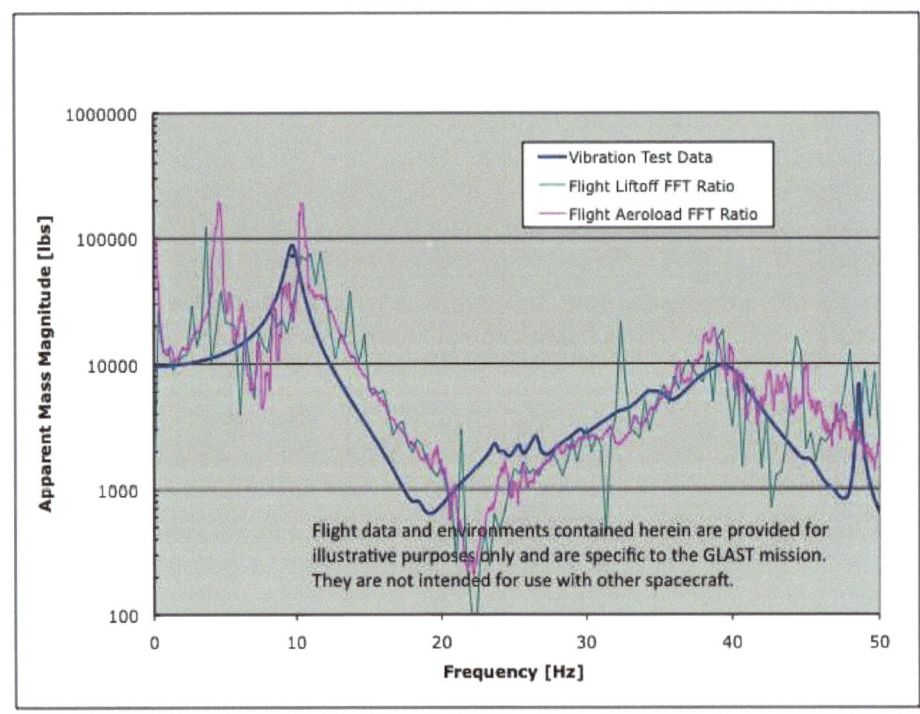

Figure 27—GLAST Spacecraft Lateral X-Axis Apparent Mass Measured in Ground Vibration Test and in Two Flight Regimes

APPROVED FOR PUBLIC RELEASE—DISTRIBUTION IS UNLIMITED

APPENDIX A

CALCULATION OF EFFECTIVE MASS

A.1 Purpose and/or Scope

The purpose of this appendix is to provide guidance.

A.2 Calculating Effective Mass

Applying the rationale of [22] and subdividing the displacement vector into unrestrained absolute displacements (u_F) and prescribed absolute displacements (u_P), the equilibrium equation is:

$$[\text{equation}] \quad \text{(Eq. A1)}$$

Let:

$$[\text{equation}] \quad \text{(Eq. A2)}$$

where:

ϕ_N are normal modes
ϕ_P are rigid body modes associated with a kinematic set of unit prescribed motions
U_N is the generalized modal relative displacement
U_P is the generalized prescribed absolute displacement.

Substituting and pre-multiplying by ϕ_N^T yields:

$$[\text{equation}] \quad \text{(Eq. A3)}$$

where:

$$M_{NN} = \phi_N^T m_{FF} \phi_N \quad \text{(Eq. A4)}$$
$$M_{NP} = \phi_N^T m_{FF} \phi_P + \phi_N^T m_{FP} I_{PP} \quad \text{(Eq. A5)}$$
$$M_{PP} = I_{PP} m_{PP} I_{PP} + I_{PP} m_{PF} \phi_P + \phi_P^T m_{FP} I_{PP} + \phi_P^T m_{FF} \phi_P \quad \text{(Eq. A6)}$$
$$F_P = I_{PP} f_P + \phi^T_P f_F \quad \text{(Eq. A7)}$$
$$F_N = \phi^T_P f_F \quad \text{(Eq. A8)}$$

For: $d^2 U_P / dt^2 = U_P = F_N = 0$, $d^2 U_n / dt^2 = -\omega_n^2 U_n$, and $U_n = 1$:
$$M_{nP}^T = -F_P / \omega_n^2 \quad \text{(Eq. A9)}$$

where:

n indicates a single mode. (Note that M_{nP}^T is in mass units.)

M_{nP}/M_{nn} is sometimes called the elastic-rigid coupling or the modal participation factor for the nth mode. If the model is restrained at a single point, the reaction (F_p) in Eq. A8 is the SPCFORCE at that point in a NASTRAN modal analysis.

The initial value of M_{PP} is the rigid body mass matrix. If a Gaussian decomposition of the total modal mass in Eq. A3 is performed, it subtracts the contribution of each normal mode, called the effective mass:

$$M_{nP}^T M_{nn}^{-1} M_{nP} \qquad (Eq.~A10)$$

from M_{PP}^n, which is the residual mass after excluding the mass associated with the already processed n modes.

Consider the ratio of the reaction force in direction p to the prescribed acceleration in direction q. The effective mass is the contribution of the nth mode to this ratio, divided by the SDOF system frequency response factor. The sum of the common-direction effective masses for all modes is equal to the total mass or moment of inertia for that direction. The effective masses are independent of the modal normalization.

NASA-HDBK-7004C

APPENDIX B

EQUATIONS FOR CALCULATING THE SIMPLE TDOF SYSTEM FORCE LIMIT

B.1 Purpose and/or Scope

The purpose of this appendix is to provide guidance.

B.2 Equations

The force limit is calculated for the TDOF system in figure 7 with different masses for the source and the payload oscillators [10]. For this TDOF system, the maximum response of the payload and, therefore, the maximum interface force occur when the uncoupled resonance frequency of the payload equals that of the source. For this case, the characteristic equation is that of a classical dynamic absorber, from [18]:

$$(\omega/\omega_o)^2 = 1 + (m_2/m_1)/2 \pm [(m_2/m_1) + (m_2/m_1)^2/4]^{0.5} \qquad \text{(Eq. B1)}$$

where:

ω is the excitation frequency
ω_o is the natural frequency of one of the uncoupled oscillators
m_1 is the mass of the source oscillator
m_2 is the mass of the load oscillator in figure 7.

The ratio of the interface force (S_{FF}) to acceleration (S_{AA}) spectral densities, divided by the magnitude squared of the payload dynamic mass (m_2), is:

$$S_{FF}/(S_{AA} m_2^2) = [1 + (\omega/\omega_o)^2/Q_2^2] / \{[1 - (\omega/\omega_o)^2]^2 + (\omega/\omega_o)^2/Q_2^2\} \qquad \text{(Eq. B2)}$$

where:

Q_2 is the quality factor of the payload.

The force spectral density, normalized by the payload mass squared and by the acceleration spectral density, at the two coupled system resonances is obtained by combining Eqs. B1 and B2. For this TDOF system, the normalized force is just slightly larger at the lower resonance frequency of Eq. B1. The maximum normalized force spectral density, obtained by evaluating Eq. B2 at the lower resonance frequency, is plotted against the ratio of payload to source mass for three values of Q_2 in figure 8.

APPROVED FOR PUBLIC RELEASE—DISTRIBUTION IS UNLIMITED

NASA-HDBK-7004C

APPENDIX C

TABLES FOR COMPLEX TDOF SYSTEM METHOD OF CALCULATING FORCE LIMITS

C.1 Purpose and/or Scope

The purpose of this appendix is to provide guidance.

C.2 Tables

Table 2, Force Limit Spectrum for Complex TDOF System with Q=50; table 3, Force Limit Spectrum for Complex TDOF System with Q=20; and table 4, Force Limit Spectrum for Complex TDOF System with Q=5, provide force limit spectra data for the complex TDOF system method of calculating force limits. Results for other Qs may be obtained either by interpolation or from the equations given in [10].

NASA-HDBK-7004C

Table 2—Force Limit Spectrum for Complex TDOF System with Q=50
(Normalized by Load Residual Mass Squared and Acceleration Spectrum)

Ratio of modal to residual mass m_1/M_1, m_2/M_2	\multicolumn{9}{c}{Residual mass ratio, M_2/M_1}								
	0.001	0.003	0.01	0.03	0.1	0.3	1	3	10
8.0, 8.0	949	950	954	966	1021	1206	1268	1261	1265
8.0, 4.0	237	237	238	239	244	260	299	270	253
8.0, 2.0	59	59	59	59	60	61	69	73	69
8.0, 1.0	15	15	15	15	15	15	17	23	22
8.0, 0.5	4	4	4	4	4	4	4	7	6
8.0, 0.25	1	1	1	1	1	1	1	2	5
8.0, 0.125	1	1	1	1	1	1	1	1	3
8.0, 0.0	1	1	1	1	1	1	1	1	1
4.0, 8.0	884	880	870	860	916	1058	1086	1134	1255
4.0, 4.0	221	221	220	219	223	253	257	253	255
4.0, 2.0	55	55	55	56	57	62	73	69	67
4.0, 1.0	14	14	14	14	15	16	22	23	22
4.0, 0.5	3	3	3	4	4	4	6	10	10
4.0, 0.25	1	1	1	1	1	1	2	5	5
4.0, 0.125	1	1	1	1	1	1	1	3	3
4.0, 0.0	1	1	1	1	1	1	1	1	1
2.0, 8.0	1640	1521	1286	1075	1003	965	996	1119	1234
2.0, 4.0	420	404	364	311	275	261	240	241	257
2.0, 2.0	106	105	99	90	80	82	70	66	63
2.0, 1.0	27	27	26	25	24	26	25	23	22
2.0, 0.5	7	7	7	7	7	9	11	10	10
2.0, 0.25	2	2	2	2	2	3	6	5	6
2.0, 0.125	1	1	1	1	1	1	3	3	4
2.0, 0.0	1	1	1	1	1	1	1	1	1
1.0, 8.0	16554	6508	2921	1510	976	909	998	1114	1212
1.0, 4.0	7333	2965	1200	583	336	249	235	240	252
1.0, 2.0	3080	1345	502	248	128	84	71	67	64
1.0, 1.0	1189	592	229	112	53	34	26	23	24
1.0, 0.5	415	245	106	51	26	16	12	11	11
1.0, 0.25	132	94	48	23	13	9	6	6	6
1.0, 0.125	39	33	21	11	7	5	4	4	4
1.0, 0.0	1	1	1	1	1	1	1	1	1
0.5, 8.0	24199	9798	3726	1761	1046	887	994	1112	1202
0.5, 4.0	10238	4417	1672	738	368	249	229	242	248
0.5, 2.0	4046	1927	747	319	143	89	72	65	65
0.5, 1.0	1454	804	335	142	62	40	27	25	23
0.5, 0.5	472	311	148	66	30	18	13	12	10
0.5, 0.25	141	110	63	31	15	9	8	7	7
0.5, 0.125	40	36	26	15	8	5	4	5	5
0.5, 0.0	1	1	1	1	1	1	1	1	1
0.25, 8.0	33910	13269	4455	1996	1026	839	955	1111	1196
0.25, 4.0	14189	6185	2155	885	393	251	227	244	247
0.25, 2.0	5342	2736	1043	405	182	96	71	66	66
0.25, 1.0	1764	1111	492	205	80	45	28	23	22
0.25, 0.5	529	396	219	104	45	23	15	12	11
0.25, 0.25	149	128	85	47	22	12	8	8	7
0.25, 0.125	41	39	31	20	11	6	5	5	5
0.25, 0.0	1	1	1	1	1	1	1	1	1
0.125, 8.0	48146	18637	6361	2411	1072	855	936	1111	1194
0.125, 4.0	19122	8823	2885	1174	411	268	230	244	246
0.125, 2.0	6642	3788	1454	508	193	106	74	67	66
0.125, 1.0	2045	1434	684	271	105	48	30	24	22
0.125, 0.5	574	477	291	139	52	27	14	13	11
0.125, 0.25	155	142	110	66	31	15	9	7	7
0.125, 0.125	41	41	36	27	16	10	6	5	5
0.125, 0.0	1	1	1	1	1	1	1	1	1
0.0, 8.0	134767	66196	13561	2836	1136	874	917	1110	1191
0.0, 4.0	37885	28769	8669	1827	464	258	233	245	245
0.0, 2.0	9820	8998	5015	1203	276	110	69	68	66
0.0, 1.0	2484	2419	1962	823	187	54	30	25	22
0.0, 0.5	625	619	580	402	111	35	17	12	11
0.0, 0.25	157	157	154	136	69	25	10	8	7
0.0, 0.125	40	40	40	38	30	14	9	5	4
0.0, 0.0	1	1	1	1	1	1	1	1	1

APPROVED FOR PUBLIC RELEASE—DISTRIBUTION IS UNLIMITED

NASA-HDBK-7004C

Table 3—Force Limit Spectrum for Complex TDOF System with Q=20
(Normalized by Load Residual Mass Squared and Acceleration Spectrum)

Ratio of modal to residual mass m1/M1, m2/M2	\multicolumn{9}{c}{Residual mass ratio, M2/M1}								
	0.001	0.003	0.01	0.03	0.1	0.3	1	3	10
8.0, 8.0	932	933	936	948	1001	1180	1240	1234	1238
8.0, 4.0	233	233	233	235	239	256	294	265	250
8.0, 2.0	58	58	58	58	59	60	68	73	68
8.0, 1.0	15	15	15	15	15	15	17	23	22
8.0, 0.5	4	4	4	4	4	4	4	7	6
8.0, 0.25	1	1	1	1	1	1	1	2	5
8.0, 0.125	1	1	1	1	1	1	1	1	3
8.0, 0.0	1	1	1	1	1	1	1	1	1
4.0, 8.0	871	867	858	849	904	1042	1067	1110	1229
4.0, 4.0	218	218	217	216	220	250	254	250	252
4.0, 2.0	55	55	55	55	56	61	72	68	67
4.0, 1.0	14	14	14	14	14	16	21	23	22
4.0, 0.5	3	3	4	4	4	4	6	10	10
4.0, 0.25	1	1	1	1	1	1	2	5	5
4.0, 0.125	1	1	1	1	1	1	1	3	3
4.0, 0.0	1	1	1	1	1	1	1	1	1
2.0, 8.0	1586	1478	1260	1061	990	946	982	1099	1201
2.0, 4.0	406	391	355	305	272	259	238	236	254
2.0, 2.0	103	101	97	88	79	82	70	65	62
2.0, 1.0	26	26	26	25	24	25	25	23	22
2.0, 0.5	7	7	7	7	7	9	10	10	10
2.0, 0.25	2	2	2	2	2	3	5	5	6
2.0, 0.125	1	1	1	1	1	1	3	3	4
2.0, 0.0	1	1	1	1	1	1	1	1	1
1.0, 8.0	11041	5731	2714	1486	967	901	984	1095	1181
1.0, 4.0	3869	2206	1105	567	332	247	233	238	248
1.0, 2.0	1228	826	432	226	125	83	71	66	64
1.0, 1.0	359	283	166	100	50	34	26	23	23
1.0, 0.5	100	89	63	42	24	15	12	11	11
1.0, 0.25	28	27	23	17	11	8	6	6	6
1.0, 0.125	8	8	8	7	5	5	4	4	4
1.0, 0.0	1	1	1	1	1	1	1	1	1
0.5, 8.0	13889	7720	3501	1726	1023	880	974	1093	1171
0.5, 4.0	4516	2895	1417	695	357	247	225	240	244
0.5, 2.0	1346	1003	561	283	136	89	70	64	65
0.5, 1.0	377	319	211	117	59	39	27	24	22
0.5, 0.5	102	95	74	48	27	17	12	11	10
0.5, 0.25	28	27	25	19	13	8	7	6	6
0.5, 0.125	8	8	8	8	6	5	4	4	4
0.5, 0.0	1	1	1	1	1	1	1	1	1
0.25, 8.0	17378	9978	4092	1944	1017	833	936	1092	1166
0.25, 4.0	5194	3725	1805	812	380	249	225	241	242
0.25, 2.0	1455	1205	741	359	173	93	71	66	65
0.25, 1.0	391	354	269	160	74	43	28	23	22
0.25, 0.5	103	99	86	63	38	22	14	12	11
0.25, 0.25	28	28	27	23	16	10	8	7	7
0.25, 0.125	8	8	8	8	7	5	5	4	4
0.25, 0.0	1	1	1	1	1	1	1	1	1
0.125, 8.0	19966	12425	5389	2331	1048	849	918	1092	1163
0.125, 4.0	5748	4417	2241	1080	400	266	228	242	241
0.125, 2.0	1533	1368	901	429	184	102	72	66	65
0.125, 1.0	400	380	312	192	91	45	29	24	22
0.125, 0.5	104	102	95	75	42	24	14	12	11
0.125, 0.25	27	28	27	26	20	13	8	7	6
0.125, 0.125	8	8	8	8	8	7	5	4	4
0.125, 0.0	1	1	1	1	1	1	1	1	1
0.0, 8.0	25114	21284	10111	2700	1125	867	900	1091	1161
0.0, 4.0	6394	6108	4156	1560	454	256	231	240	240
0.0, 2.0	1608	1590	1409	757	257	109	68	67	66
0.0, 1.0	404	403	390	310	148	52	30	25	22
0.0, 0.5	102	102	101	95	60	30	16	12	11
0.0, 0.25	27	27	27	26	22	17	9	7	6
0.0, 0.125	8	8	8	7	7	6	6	5	4
0.0, 0.0	1	1	1	1	1	1	1	1	1

APPROVED FOR PUBLIC RELEASE—DISTRIBUTION IS UNLIMITED

NASA-HDBK-7004C

Table 4—Force Limit Spectrum for Complex TDOF System with Q=5
(Normalized by Load Residual Mass Squared and Acceleration Spectrum)

Ratio of modal to residual mass m1/M1, m2/M2	Residual mass ratio, M2/M1								
	0.001	0.003	0.01	0.03	0.1	0.3	1	3	10
8.0, 8.0	702	703	706	715	752	866	865	876	873
8.0, 4.0	177	178	178	179	183	197	221	212	205
8.0, 2.0	46	46	46	46	46	48	56	58	55
8.0, 1.0	12	12	12	12	12	13	14	19	18
8.0, 0.5	4	4	4	4	4	4	4	6	5
8.0, 0.25	1	1	1	1	1	1	1	2	4
8.0, 0.125	1	1	1	1	1	1	1	1	2
8.0, 0.0	1	1	1	1	1	1	1	1	1
4.0, 8.0	687	687	685	689	739	803	827	838	841
4.0, 4.0	174	174	174	175	182	207	205	203	198
4.0, 2.0	45	45	45	45	47	52	59	55	53
4.0, 1.0	12	12	12	12	13	14	19	19	18
4.0, 0.5	3	3	3	4	4	4	6	7	7
4.0, 0.25	1	1	1	1	1	1	2	4	4
4.0, 0.125	1	1	1	1	1	1	1	2	2
4.0, 0.0	1	1	1	1	1	1	1	1	1
2.0, 8.0	1006	983	927	860	808	758	786	839	826
2.0, 4.0	256	254	247	235	228	209	197	198	192
2.0, 2.0	66	66	65	64	64	64	57	56	56
2.0, 1.0	18	18	18	18	19	21	20	19	19
2.0, 0.5	5	5	5	5	6	7	8	8	8
2.0, 0.25	2	2	2	2	2	3	4	4	4
2.0, 0.125	1	1	1	1	1	1	2	2	2
2.0, 0.0	1	1	1	1	1	1	1	1	1
1.0, 8.0	1700	1618	1389	1048	787	721	758	835	826
1.0, 4.0	434	430	397	324	244	202	191	196	193
1.0, 2.0	113	113	111	102	80	66	58	55	54
1.0, 1.0	30	31	31	31	28	24	20	20	19
1.0, 0.5	9	9	10	10	10	9	8	8	8
1.0, 0.25	3	3	4	4	4	4	4	4	4
1.0, 0.125	2	2	2	2	2	2	2	2	2
1.0, 0.0	1	1	1	1	1	1	1	1	1
0.5, 8.0	1703	1655	1456	1147	831	705	738	821	826
0.5, 4.0	433	432	409	350	263	205	193	190	193
0.5, 2.0	112	113	112	104	85	67	59	55	53
0.5, 1.0	31	31	31	31	28	25	20	19	19
0.5, 0.5	10	10	10	10	10	9	9	8	8
0.5, 0.25	4	4	4	4	4	4	4	4	4
0.5, 0.125	2	2	2	2	2	2	2	2	2
0.5, 0.0	1	1	1	1	1	1	1	1	1
0.25, 8.0	1698	1655	1523	1181	828	730	742	814	826
0.25, 4.0	434	431	414	376	280	209	192	187	194
0.25, 2.0	112	113	112	107	94	73	57	54	53
0.25, 1.0	31	31	31	31	29	26	21	20	19
0.25, 0.5	10	10	10	10	10	9	9	8	8
0.25, 0.25	4	4	4	4	4	4	4	4	4
0.25, 0.125	2	2	2	2	2	2	2	2	2
0.25, 0.0	1	1	1	1	1	1	1	1	1
0.125, 8.0	1705	1692	1638	1291	869	713	743	810	826
0.125, 4.0	433	433	429	385	292	211	188	187	194
0.125, 2.0	112	112	113	112	97	74	59	53	53
0.125, 1.0	31	31	31	31	31	27	22	19	19
0.125, 0.5	10	10	10	10	10	10	9	8	8
0.125, 0.25	4	4	4	4	4	4	4	4	4
0.125, 0.125	2	2	2	2	2	2	2	2	2
0.125, 0.0	1	1	1	1	1	1	1	1	1
0.0, 8.0	1681	1681	1655	1435	862	692	745	807	826
0.0, 4.0	425	425	425	424	310	215	184	187	194
0.0, 2.0	111	111	111	111	100	74	61	54	52
0.0, 1.0	31	31	31	31	31	29	22	19	19
0.0, 0.5	10	10	10	10	10	10	9	9	8
0.0, 0.25	4	4	4	4	4	4	4	4	4
0.0, 0.125	2	2	2	2	2	2	2	2	2
0.0, 0.0	1	1	1	1	1	1	1	1	1

APPROVED FOR PUBLIC RELEASE—DISTRIBUTION IS UNLIMITED

APPENDIX D

SWEITZER'S NOTCHING CRITERION

D.1 Purpose and/or Scope

The purpose of this appendix is to provide guidance.

D.2 Sweitzer's Notching Criterion

In 1987, Sweitzer proposed a method for notching the input acceleration specification at lightly damped resonances of the test item to reduce the overtesting [36] associated with the very high mechanical impedance of vibration test machines (shakers). Although Sweitzer's method does not directly involve force limiting, it is discussed here because his notching criterion does provide a simple guideline (see section 6 of this Handbook) for assessing the notch depth resulting from force limiting at lightly damped resonances. Sweitzer's method is heuristic and difficult to justify because he did not quantitatively relate his notching criterion to the mechanical impedance of the flight mounting structure. However, it is shown here that Sweitzer's notching criterion can be associated with specific ratios of the load to source masses in the simple TDOF system model discussed in section 5.4 of this Handbook.

For a random vibration test, Sweitzer's method involves reducing (notching) the PSD of the input acceleration specification ($S_{AA}(f)$) in the vicinity of the fundamental resonance frequency (f_o) according to:

$$S'_{AA}(f) = S_{AA}(f)/|H(f)_{r/a}|; \quad 0 < f < 2^{0.5} f_o \qquad (Eq.\ D1)$$

where:

$S'_{AA}(f)$ is the PSD of the notched input acceleration
$H(f)_{r/a}$ is the FRF of the ratio of an acceleration response to that of the acceleration input.

(For a sine vibration test, the PSDs in Eq. D1 are replaced by their respective amplitudes, and the FRF in the denominator is replaced by its square root.) In practice, the FRF in Eq. D1 would be measured in a preliminary low-level vibration test, and some judgment would be required to select a representative response measurement position.

For the simple TDOF system in figure 7 in section 5.4 of this Handbook, the FRF of the ratio of the load acceleration response amplitude to the load acceleration input at the load resonance frequency is Q, the load quality factor. So applying Eq. D1 to a random vibration test of the load, the input acceleration PSD would be notched by a factor of Q at the load resonance frequency. As the ratio of the load acceleration response PSD to the load acceleration input PSD is Q^2, implementing Sweitzer's notching criterion would result in a net amplification in the acceleration response PSD of only Q. (The net amplification in a sine vibration test would be reduced to $Q^{0.5}$.) Another way of saying this for both random and sine vibration tests is that Sweitzer's notching criterion reduces the amplification at resonance to half its original value in decibels. (For example, a 20-dB resonance becomes a 10-dB resonance, etc.)

Comparing the foregoing result with Eq. 2 in section 5.3 of this Handbook, it may be seen that Sweitzer's notching criterion is equivalent to choosing a C^2 equal to Q in the semi-empirical method of force limiting. Thus, referring to figure 8 in section 5.4 of this Handbook, in which C^2 is the ordinate, Sweitzer's notching criterion corresponds to a load to source mass ratio in the simple TDOF system method of approximately $M_2/M_1 = 0.3$ for Q =5, $M_2/M_1 = 0.06$ for Q =20, and $M_2/M_1 = 0.02$ for Q =50. This comparison indicates that Sweitzer's notching criterion is reasonable for relatively heavy, lightly damped test items, but any notching should be supported by a consideration of the relative magnitudes of the test item and mounting structure impedances for the problem at hand.